**DMV Seminar
Band 27**

Classical Nonintegrability, Quantum Chaos

With a contribution by Viviane Baladi

Andreas Knauf
Yakov G. Sinai

Birkhäuser Verlag
Basel · Boston · Berlin

Authors:

Andreas Knauf
Fachbereich 3 - Mathematik, MA 7-2
Technische Universität Berlin
Straße des 17. Juni 135
10623 Berlin
Germany
e-mail: Knauf@math.tu-berlin.de

Yakov G. Sinai
Department of Mathematics
Princeton University
Princeton, NJ 08544
USA
e-mail: sinai@math.princeton.edu

Viviane Baladi
Section de Mathématiques
Université de Genève
CP 240
1211 Genève 24
Switzerland
e-mail: baladi@sc2a.unige.ch

1991 Mathematical Subject Classification 81Q50, 81U99, 58F11

A CIP catalogue record for this book is available from the Library of Congress, Washington D.C., USA

Deutsche Bibliothek Cataloging-in-Publication Data
Knauf, Andreas:
Classical nonintegrability, quantum chaos / Andreas Knauf ;
Yakov G. Sinai. With a contribution by Viviane Baladi. - Basel
; Boston ; Berlin : Birkhäuser, 1997
   (DMV-Seminar ; Bd. 27)
   ISBN 3-7643-5708-8 (Basel ...)
   ISBN 0-8176-5708-8 (Boston)
NE: Sinaj, Jakov G.:; Deutsche Mathematiker-Vereinigung: DMV-
   Seminar

© 1997 Birkhäuser Verlag, P.O. Box 133, CH-4010 Basel, Switzerland
Camera-ready copy prepared by the author
Printed on acid-free paper produced from chlorine-free pulp. TCF ∞
Cover design: Heinz Hiltbrunner, Basel
Printed in Germany
ISBN 3-7643-5708-8
ISBN 0-8176-5708-8

Dynamics in $\mathbb{R}P^2$

Contents

Chapter 1

Introduction

Our DMV Seminar on 'Classical Nonintegrability, Quantum Chaos' intended to introduce students and beginning researchers to the techniques applied in nonintegrable classical and quantum dynamics.

Several of these lectures are collected in this volume.

The basic phenomenon of nonlinear dynamics is mixing in phase space, leading to a positive dynamical entropy and a loss of information about the initial state. The nonlinear motion in phase space gives rise to a linear action on phase space functions which in the case of iterated maps is given by a so-called transfer operator.

Good mixing rates lead to a spectral gap for this operator. Similar to the use made of the Riemann zeta function in the investigation of the prime numbers, dynamical zeta functions are now being applied in nonlinear dynamics.

In Chapter 2 V. Baladi first introduces dynamical zeta functions and transfer operators, illustrating and motivating these notions with a simple one-dimensional dynamical system.

Then she presents a commented list of useful references, helping the newcomer to enter smoothly into this fast-developing field of research.

Chapter 3 on irregular scattering and Chapter 4 on quantum chaos by A. Knauf deal with solutions of the Hamilton and the Schrödinger equation. Scattering by a potential force tends to be irregular if three or more scattering centres are present, and a typical phenomenon is the occurrence of a Cantor set of bounded orbits. The presence of this set influences those scattering orbits which come close. Since almost all orbits escape to infinity, the description using symbolic dynamics is simpler than for bounded motion.

By definition quantum chaos is the quantal motion of a classically nonintegrable system. But the classical phase space structures are blurred by the uncertainty principle and are only visible in the semiclassical limit of Planck's constant $\hbar \searrow 0$. Then in a sense made precise by the Schnirelman theorem most eigenfunctions of a classically chaotic Hamiltonian are equidistributed over the energy shell.

1

Chapters 5, 6 and 7 are lectures by Ya. Sinai. In the first of these lectures the hierarchy of ergodic, mixing and $\mathbb{K}$-systems is presented, starting with Birkhoff's theorem on time-averages. Then the impact of these classical notions on modern research topics is sketched.

Chapter 6 deals with expanding maps which by definition have the property to locally increase distances. These maps have an absolutely continuous invariant probability measure, and a complete proof of that fact is presented for the one-dimensional case. Maybe even more important than the mere fact is the method of proof which connects these discrete dynamical systems with spin chains of statistical mechanics.

The last chapter on Liouville surfaces considers the spectral properties of the Laplace-Beltrami operator on tori with a class of metrics that makes the geodesic flow integrable. A very precise control of the spectral statistics is achieved.

There were several additional talks by participants working in the field which enriched the seminar.

Acknowledgements: V. B. is partially supported by the Fonds National Suisse de la Recherche Scientifique. A. K. thanks Arnd Bäcker and Nicolas Friese for useful comments. We all thank Prof. M. Kreck and the staff of Oberwolfach for their support which made the stay so pleasurable.

V. Baladi, A. Knauf, Ya. Sinai

Chapter 2

A Brief Introduction to Dynamical Zeta Functions

2.1 Introduction and Motivation

2.1.1 Transfer Operators

We shall take the point of view that dynamical zeta functions are useful objects to describe the spectrum of transfer operators. To define a *transfer operator*, we use two ingredients:

- a map $f : X \to X$ of a topological or metric space X to itself (the *dynamical system*), with the property that $f^{-1}(x)$ is an at most countable set for each $x \in X$;

- a *weight* $g : X \to \mathbb{C}$.

In order to get interesting results, one usually requires additional assumptions: e.g., the map f is supposed to be locally expanding or hyperbolic, and both the map f and the weight g should satisfy some smoothness condition (for example Hölder or Lipschitz continuity if X is a metric space, differentiability or analyticity if X has a manifold structure).

A *transfer operator* is a linear operator $\mathcal{L}_g$ which acts on a suitable vector space of functions $\varphi : X \to \mathbb{C}$ (e.g., the Banach space of bounded functions if $\sum_{y \in f^{-1}(x)} |g(y)| < \infty$ for all x) according to the formula:

$$\mathcal{L}_g \varphi(x) = \sum_{y \in f^{-1}(x)} \varphi(y) g(y) \,. \tag{2.1}$$

Before being used in the framework of dynamical systems, transfer operators appeared in statistical mechanics (see the classical treatise on thermodynamic formalism [31]).

3

Example 1. Let us take $X = I = [0,1]$ the unit interval and $f(x) = 2x \,(\text{mod}\,1)$. We consider first the unweighted case, i.e., we set $g \equiv 1$. Then it is easy to see that the two-dimensional vector space V_2 of functions $\varphi : I \to \mathbb{C}$ (viewing as equivalent two functions which differ on an at most countable set) such that $\varphi(x) = \varphi_L \in \mathbb{C}$ for $0 \leq x < 1/2$, and $\varphi(x) = \varphi_R \in \mathbb{C}$ for $1/2 \leq x \leq 1$, is preserved by $\mathcal{L}_g$. In the basis just described for V_2, one checks that the matrix of $\mathcal{L}_g$ is simply $A = \begin{pmatrix} 1 & 1 \\ 1 & 1 \end{pmatrix}$, the eigenvalues of which are 0 and 2. Now we make the following trivial observation: since $\operatorname{tr} A^n = 2^n$ is exactly the cardinality of the set $\operatorname{Fix} f^n := \{x \in I \mid f^n(x) = x\}$ of fixed points of f^n, the unweighted zeta function defined formally by the power series

$$\zeta(z) = \exp \sum_{n \geq 1} \frac{z^n}{n} \#\operatorname{Fix} f^n \tag{2.2}$$

satisfies

$$\zeta(z) = \exp \sum_{n \geq 1} \frac{z^n}{n} \operatorname{tr} A^n = \frac{1}{\det(\hat{\mathbb{1}} - zA)} \tag{2.3}$$

(where $\hat{\mathbb{1}}$ is just the two by two identity matrix).

2.1.2 Invariant Function Spaces

In our first, rather trivial, example the (unweighted) dynamical zeta function (2.2) was a rational function with poles the inverses of the eigenvalues of the transfer operator (2.1) acting on a finite dimensional vector space. For dynamical systems having a finite Markov grammar (e.g. Axiom A diffeomorphisms), this result (in particular the formula (2.3), with $\#\operatorname{Fix} f^n$ replaced by $\sum_{x \in \operatorname{Fix} f^n} \prod_{k=0}^{n-1} g(f^k(x))$) may be generalized *whenever the weight g is constant or locally constant* (see e.g. [20]). However, for non-constant weights there is in general no obvious finite dimensional space preserved by the transfer operator. One needs to find an adequate invariant Banach space (when possible a Hilbert space). It turns out, unfortunately, that the transfer operator acting on such a Banach space, although bounded, is usually not compact. We illustrate this remark with our second (and last) example, which shows also why it is natural to consider non-constant weights.

Example 2.a. Again we take $X = I$, which we assume to be partitioned into two sub-intervals $L = [0, c[$ and $R = [c, 1]$. We consider now a function $f : I \to I$ such that the restrictions of f to L and R are monotone and C^2, with $f(R) = I$, and such that f has a C^2 extension $\bar{f}$ to $\bar{L}$ with $\bar{f}(\bar{L}) = I$. Finally, we impose the following important *locally expanding* condition: there exists $\lambda > 1$ with $|\bar{f}'| \geq \lambda$ on $\bar{L}$ and $|f'| \geq \lambda$ on R. (From now on we neglect in this example all difficulties related to the boundary point c.) Now, if we take $g = 1/|f'|$ for our weight, we obtain the key property that Lebesgue measure dx on I is preserved by the dual of $\mathcal{L}_g$. By definition this means that for any $\varphi \in L^1(dx)$ we have

$$\int \varphi(x)\, dx = \int \mathcal{L}_g(\varphi)(x)\, dx \qquad (2.4)$$

(just use the change of variable formula in an integral), which we abbreviate as $\mathcal{L}_g^*(dx) = dx$. It follows that if we find a positive fixed point $\varphi_0 \in L^1(dx)$ for $\mathcal{L}_g$, then we may construct an absolutely continuous f-invariant measure by setting $d\mu_0 = \varphi_0\, dx$. Indeed, for any bounded φ we get by using (2.4) that

$$
\begin{aligned}
\int \varphi(f(x))\, \varphi_0(x)\, dx &= \int \mathcal{L}_g((\varphi \circ f)\, \varphi_0)(x)\, dx \\
&= \int \varphi(x) \mathcal{L}_g(\varphi_0)(x)\, dx \qquad (2.5) \\
&= \int \varphi(x) \varphi_0(x)\, dx\,.
\end{aligned}
$$

(In the second equality of (2.6) we just used formula (2.1) for $\mathcal{L}_g$.) It is therefore natural to try to understand the spectrum and eigenfunctions of $\mathcal{L}_g$ acting on a suitable space for $g = 1/|f'|$. Clearly, finite-dimensional spaces of locally constant functions, as the vector space V_2 in Example 1, will not be preserved in general. Unfortunately, the Banach space $L^1(dx)$, although invariant under $\mathcal{L}_g$, is "too big:" the spectrum of $\mathcal{L}_g$ acting on $L^1(dx)$ consists in the entire closed unit disc (each point of the open disk is actually an eigenvalue of infinite multiplicity, see e.g. [44]). It turns out that the spectrum of $\mathcal{L}_g$ acting on the Banach space of continuous functions (with supremum norm) is also the entire unit disc. However, by combining the contraction property of the two inverse branches of f with the smoothness of f and g, we will see below that $\mathcal{L}_g$ preserves the "smaller" Banach space $C^1(I)$ of C^1 functions (endowed with the norm $\sup|\varphi| + \sup|\varphi'|$) and that its spectrum for this space has a *gap*. (We shall explain how this spectral property is useful to show exponential mixing properties for differentiable observables.)

2.1.3 Quasicompactness

For L a bounded linear operator acting on a Banach space B, we define the *essential spectral radius* of L to be the smallest nonnegative number ρ such that the spectrum of L outside of the disc $\{z \in \mathbb{C} \mid |z| \le \rho\}$ consists of isolated eigenvalues of finite multiplicity. (This definition does not exclude the case where these eigenvalues accumulate on the circle $|z| = \rho$.)

In many interesting situations, including the framework of Example 2, one can find a Banach space B of functions $\varphi : X \to \mathbb{C}$ which is invariant under $\mathcal{L}_g$ and such that:

- there is an upper bound R for the spectral radius of $\mathcal{L}_g : B \to B$ and an upper bound $R_{\mathrm{ess}} < R$ for the essential spectral radius of $\mathcal{L}_g : B \to B$;

- if one assumes additionally that the weight g is real-valued and strictly positive, then $R > 0$ is actually an eigenvalue of $\mathcal{L}_g$, with a positive eigenfunction φ_0, and a positive eigenfunctional $d\nu_0$ for $\mathcal{L}_g^*$ which is a Borel measure. (In particular, a trivial modification of the algebra in (2.6) implies that $d\mu_0 = \varphi_0 d\nu_0$ is an f-invariant finite measure.)

For positive weights, it is often possible to prove under additional assumptions that the real positive eigenvalue R is simple and is the only eigenvalue of modulus R. In this case, one says that the transfer operator has a *spectral gap*, and one has

$$\tau := \sup\{|z| \mid z \in \text{spectrum}\, \mathcal{L}_g , z \neq R\} < R. \tag{2.6}$$

Example 2.b. In the setting of Example 2.a, one can prove that the spectral radius of $\mathcal{L}_{1/|f'|}$ acting on $C^1(I)$ is equal to 1 and that the essential spectral radius is not larger than $1/\lambda < 1$ (see [44] for a better bound). Also, one sees that there is a positive fixed function $\varphi_0 \in C^1(I)$. (The fixed eigenfunctional $d\nu_0$ is just Lebesgue measure dx as already mentioned.) Before discussing this example further, we note that a crucial bound in obtaining these results is the following inequality: for any $1 < \Lambda < \lambda$, there is a constant $C > 0$ so that for all $\varphi \in C^1(I)$ and all $n \geq 1$:

$$\sup\left|\frac{d}{dx}(\mathcal{L}_g^n \varphi)(x)\right| \leq \frac{C}{\Lambda^n} \sup\left|\frac{d}{dx}\varphi(x)\right| + C \sup|\varphi|. \tag{2.7}$$

(See e.g. [58] for various occurrences of this bound.) To prove (2.7), expand

$$\mathcal{L}_g^n \varphi(x) = \sum_{y:f^n(y)=x} \frac{\varphi(y)}{|(f^n)'(y)|}, \tag{2.8}$$

and differentiate the right-hand-side of (2.8) with respect to the variable x, applying the Leibniz formula to each summand. The contraction factor $1/\Lambda^n$ comes from the interior derivative which appears when differentiating $\varphi(y)$ with respect to x. The proof of (2.7) is a simple occurrence of a general phenomenon: the two building blocks of transfer operators are *composition* (here, with contracting local inverse branches) and *multiplication* by a weight, so that change of variables and the Leibniz formula (or integration by parts) are the key tools used to obtain bounds.

In fact, one may also show in the situation of Example 2.a that the eigenvalue 1 is simple and is the only eigenvalue of modulus 1. The simplicity of the eigenvalue is related to the fact that $d\mu_0 = \varphi_0\, dx$ is the unique absolutely continuous invariant measure of f and that $(f, d\mu_0)$ is ergodic. The existence of the gap (i.e., the fact that $\tau < 1$ in the notation of (2.6)) is linked to mixing properties of $(f, d\mu_0)$. Specifically, for any $1 > \tilde{\tau} > \tau$ there is a constant $C > 0$ so that if $\psi_1 \in L^1(dx)$ and $\psi_2 \in C^1(I)$, the *correlation function*

$$C_{\psi_1,\psi_2}(k) = \int \psi_1 \circ f^k\, \psi_2\, d\mu_0 - \int \psi_1\, d\mu_0 \int \psi_2\, d\mu_0 \tag{2.9}$$

satisfies

$$
\begin{aligned}
|C_{\psi_1,\psi_2}(k)| &= \left| \int \left[\mathcal{L}_g^k (\psi_1 \circ f^k \, \psi_2 \, \varphi_0) - \left(\int \psi_2 \, d\mu_0 \right) \psi_1 \, \varphi_0 \right] dx \right| \\
&= \left| \int \left[\mathcal{L}_g^k (\psi_2 \, \varphi_0) - \left(\int \psi_2 \, d\mu_0 \right) \varphi_0 \right] \psi_1 dx \right| \\
&\leq \sup \left| \mathcal{L}_g^k (\psi_2 \, \varphi_0) - \varphi_0 \cdot \int (\psi_2 \varphi_0) \, dx) \right| \int |\psi_1| \, dx \\
&\leq C \, \tilde{\tau}^n \left(\sup |\psi_2| + \sup |\psi_2'| \right) \int |\psi_1| \, dx .
\end{aligned}
\tag{2.10}
$$

Therefore the correlation functions associated to the unique absolutely continuous invariant measure $d\mu_0$ and C^1 observables decay exponentially fast (with uniform rate τ). In the last inequality of (2.10) we used the spectral decomposition of $\mathcal{L}_g$ given by

$$
\{ |z| \mid z \in \text{spectrum} \, \mathcal{L}_g , |z| < 1 \} \cup \{1\}.
$$

(We refer e.g. to [58] for more details.)

As a last comment, we mention that whenever the map f is piecewise C^r (for some $r \geq 3$), then the weight $g = 1/|f'|$ is piecewise C^{r-1}, and the transfer operator acts on the Banach space $C^{r-1}(I)$. One can prove that $\varphi_0 \in C^{r-1}(I)$ and that the essential spectral radius of $\mathcal{L}_g$ is not larger than $1/\lambda^{r-1}$. When f is piecewise analytic, the operator $\mathcal{L}_g$ is a compact operator when acting on holomorphic functions on a complex neighbourhood of I (with a bounded extension to the boundary, and using the supremum norm). We refer to [43] for the analytic case, and to [47], [48] for the differentiable case where locally expanding maps on more general manifolds are considered (with $g = 1/|\det Df|$ or more general smooth weights).

2.1.4 Weighted Dynamical Zeta Functions

We return again to the very general framework of a transfer operator $\mathcal{L}_g$ constructed from a dynamical system f and a weight g. Having fixed a Banach space $\mathcal{B}$ for which both claims of the subsection 2.1.3 on quasicompactness can be proved, our aim is to find, for various relevant classes of f, g, a suitable definition of a (generalized) trace "tr" $\mathcal{L}_g$ for $\mathcal{L}_g$ (and all its powers). Recall that $\mathcal{L}_g$ acting on $\mathcal{B}$ will *not* be a compact operator in general. In particular, we shall *not* require that this generalized trace be related to the sum of the eigenvalues of $\mathcal{L}_g$ (which might form an uncountable set). Instead, our wish is that the *generalized Fredholm determinant*, defined formally by

$$
d_g(z) = \exp - \sum_{n \geq 1} \frac{z^n}{n} \, \text{"tr"} \, \mathcal{L}_g^n ,
\tag{2.11}
$$

has the properties that:

- $d_g(z)$ defines an analytic function in a disc $|z| < R_{\text{ess}}^{-1}$, where R_{ess} is an upper bound for the essential spectral radius of $\mathcal{L}_g : \mathcal{B} \to \mathcal{B}$;

- the zeroes of $d_g(z)$ in this disc are exactly the inverses of the eigenvalues of $\mathcal{L}_g : \mathcal{B} \to \mathcal{B}$ in the corresponding annulus (the order of the zero coinciding with the algebraic multiplicity of the eigenvalue).

If the two above properties hold, the function $d_g(z)$ clearly deserves to be called a generalized Fredholm determinant. We shall see that sometimes the function $\zeta_g(z) = 1/d_g(z)$ has the form

$$\zeta_g(z) = \exp \sum_{n \geq 1} \frac{z^n}{n} \sum_{x \in \text{Fix } f^n} \prod_{k=0}^{n-1} g(f^k(x)) \qquad (2.12)$$

which is a general expression for the *dynamical zeta function* associated to the dynamics f and the weight g. Note that there are variants for the definition of $\zeta_g(z)$, in particular when the number of elements in Fix f^n can be infinite (even uncountable, see [71, 72]).

We now illustrate some possible definitions in the simple setting of Example 2.a.

Example 2.c. We consider again our interval map f and the weight $g = 1/|f'|$, assuming that f is piecewise C^r for some $r \geq 2$. There are basically two ways to define the trace in this framework:
a) The *counting trace* is defined by taking

$$\text{tr}^c \mathcal{L}_g = \sum_{y \in \text{Fix } f} g(y) \qquad (2.13)$$

(so that $\text{tr}^c \mathcal{L}_g^n = \sum_{y \in \text{Fix } f^n} \prod_{k=0}^{n-1} g(f^k y)$, note that the number of terms in the sum is finite but grows exponentially in n). With this definition, we just get $d_g^c(z) = 1/\zeta_g(z)$ with ζ_g defined in (2.12), and one may prove ([48]):

Theorem 2.1 *The power series for $\zeta_g(z)$ defines an analytic function in the disc of radius 1, with a meromorphic extension to the disc of radius λ where its poles are exactly the inverses of the eigenvalues of $\mathcal{L}_g : C^{r-1}(I) \to C^{r-1}(I)$ of modulus larger than $1/\lambda$ (with correct multiplicity).*

b) The *flat trace* is defined by setting

$$\text{tr}^b \mathcal{L}_g = \sum_{y \in \text{Fix } f} \frac{g(y)}{(1 - 1/|f'(y)|)} \qquad (2.14)$$

(and thus $\text{tr}^b \mathcal{L}_g^n = \sum_{y \in \text{Fix } f^n} (\prod_{k=0}^{n-1} g(f^k y))/(1 - 1/|(f^n)'(y)|)$). Using this definition of the trace, we get a determinant $d_g^b(z)$ such that [48]:

Theorem 2.2 *The power series for $d_g^b(z)$ is analytic in the disc of radius λ^{r-1} where its zeroes are exactly the inverses of the eigenvalues of $\mathcal{L}_g : C^{r-1}(I) \to C^{r-1}(I)$ of modulus larger than $1/\lambda^{r-1}$ (with correct multiplicity).*

The flat determinant therefore "sees" more of the spectrum of $\mathcal{L}_g$ than the counting determinant (or zeta function) whenever $r - 1 > 1$. We refer e.g. to the survey [8] for a heuristic justification of the formula for the flat trace and references to papers of Atiyah and Bott which inspired the terminology.

Versions of Theorem 1 and Theorem 2 on the counting trace and the flat trace stated in Example 2.c apply to any locally λ-expanding C^r map $f : X \to X$ and C^{r-1} weight g (see [47], [48]). The spectral radius is then bounded by the exponential of the topological pressure of $\log |g|$.

2.2 Commented Bibliography

The bibliography that we have assembled here is certainly not complete nor systematic. Although it clearly reflects the tastes (and sometimes the ignorance) of the author, we hope that it can serve as a useful introduction to the reader who wishes to enter the field. We have divided the list of references into several sublists, hoping to make it more accessible. Many topics have been completely omitted, for example random dynamical systems. Finally, we have for the sake of conciseness stated most results in a rather vague and intuitive way: we refer to the cited papers (or to the surveys [7, 8, 9, 10, 11, 12, 13]) for precise formulations.

2.2.0 Foundations

The first sublist contains on the one hand two of the earliest references to dynamical zeta functions in the literature ([1], [6]) and on the other a few useful books in functional analysis ([2] and [5] contain general background, and [3], [4] specialized topics useful in particular to read the references in sublists 4.A, 5.A and 7). We have not attempted to list any elementary books on dynamical systems or ergodic theory.

2.2.1 Surveys

This list should also include the book of Parry and Pollicott [28] and the first chapter of Ruelle's recent book [66] which contains a pleasant and broad-viewed introduction to dynamical zeta functions. Reference [11] is a beautiful and comprehensive account of the physicist's viewpoint, and contains many applications, in particular to quantum chaos. Reference [7] contains an elementary exposition of Parry and Pollicott's "prime-orbit theorem" for Axiom A flows. Surveys [8] and [13] summarize many of the known results. We plan to include the more recent rigorous breakthroughs in [9].

2.2.2 Applications

There should be many more references here! We have mainly selected a striking application to Feigenbaum period-doubling ([15], [18]) and a link with Riemann zeta functions [17].

2.2.3 Subshifts of Finite Type and Axiom A

This subsection is the longest one and could actually be much longer: we have abstained from trying to give references to the original papers (in particular by Sinai, Ruelle, Bowen, Ratner) establishing the basic results in the ergodic theory of hyperbolic diffeomorphisms and flows, since there exist some very good books describing this material (notably [19], [28], and [31]). However, we have listed a few original papers specific to dynamical zeta functions, although the contents of some of them have been presented in [28].

References [19], [21] contain the basic theory of Axiom A diffeomorphisms and flows, showing how many of their ergodic properties can be studied (via Markov partitions and symbolic dynamics) by understanding subshifts of finite type (and their suspensions under Lipschitz or Hölder return times) with Lipschitz or Hölder weights. The observation of Bowen and Lanford that the unweighted zeta function of a subshift of finite type is rational is contained in [20], and the application by Manning to Axiom A diffeomorphisms (whose unweighted zeta functions (2.2) are also rational) appeared in [27].

A succession of results on transfer operators and zeta functions for subshifts of finite type f and Lipschitz (or Hölder) weights g is contained in [29], [30] (Pollicott), [33], [34] (Ruelle), and [25], [26] (Haydn). Many of them are collected in the book [28]. They give bounds on the spectral radius and essential spectral radius of the transfer operator (2.1) acting on Lipschitz functions, and say that the weighted zeta function (2.12) is meromorphic in a disc where its poles are the inverse eigenvalues of $\mathcal{L}_g$ in the corresponding annulus. The unweighted zeta function of a flow with countably many closed orbits is defined formally by

$$\zeta^*(s) = \prod_{\tau}(1 - e^{-s\ell(\tau)})^{-1} \qquad (2.15)$$

(where the product is over all closed orbits τ, with primitive length $\ell(\tau)$). The zeta functions $\zeta^*(s)$ of flows obtained by suspending subshifts of finite type under a Lipschitz return time r are shown to be analytic in a half-plane $\Re s > \eta$ (η being actually the topological entropy of the flow), and meromorphic in a larger half-plane $\Re s > \delta$ (the poles there corresponding to values of s such that the operator $\mathcal{L}_{e^{-sr}}$ weighted with $g = \exp(-sr)$ has the number 1 in its spectrum). Simplified sketches of some of the proofs may be found in the review [8].

Gallavotti [24] constructed examples of suspensions under Lipschitz return times where a non-polar singularity was present arbitrarily close to the vertical $\Re s = \delta$. Ruelle [32] and Pollicott constructed examples of mixing suspensions for which poles accumulated on the vertical $\Re s = \eta$, showing that topologically mixing Axiom A flows need not have exponential decay of correlations for the measure of maximal entropy (or other Gibbs measures).

Very recently, Dolgopyat [23] found sufficient conditions under which the poles of a mixing Axiom A flow cannot accumulate on the verical $\Re s = \eta$, and showed that generically the accumulation of these poles when it occurs cannot be "too fast," ensuring rapid decay (in the sense of Schwartz) of correlations for Gibbs measures associated to Lipschitz interactions (in particular the SRB measure). He also showed [22] that in the case of Anosov flows with C^1 stable and unstable foliations (in particular the geodesic flows on surfaces of negative curvature) the conditions forbidding the accumulation of poles were satisfied.

2.2.4 The Smooth Expanding Case

4.A. Analytic expanding case

Over 20 years, ago Ruelle [43] observed that Grothendieck's theory of nuclear operators [3, 4] could be applied to transfer operators when both the dynamics and the weight were analytic (assuming that the dynamics is uniformly locally expanding). In this case, the operators (acting on a suitable space of bounded holomorphic functions) are compact, and the flat trace

$$\operatorname{tr}^b \mathcal{L}_g = \sum_{x \in \operatorname{Fix}(x)} \frac{g(x)}{\operatorname{Det}\left(1 - Df^{-1}(x)\right)} \tag{2.16}$$

is actually the sum of the eigenvalues of $\mathcal{L}_g$. The corresponding determinant $d_g^b(z)$ is an entire function with zeroes the inverses of the eigenvalues of $\mathcal{L}_g$. Ruelle applied these powerful results to Anosov diffeomorphisms or flows, under the strong assumption that their stable/unstable foliations are analytic. This assumption was required because the strategy then was to reduce to the expanding situation by projecting along stable manifolds. (It is unfortunately very rarely satisfied: the invariant foliations are usually only Hölder.) Extensions of these results (in particular a correction of the asymptotics of the eigenvalues) were obtained by Fried [36].

Mayer [40, 41] used this approach to study the Gauss map

$$x \mapsto \{1/x\} \qquad (0 < x \leq 1).$$

One striking result in [40] is the proof of the reality of the spectrum of some weighted transfer operators associated to the Gauss map: this is obtained by considering the action of the operator on suitable Hilbert spaces of Hardy functions and proving self-adjointness. The survey [42] discusses some applications.

Papers [35] and [37, 38, 39] contain some rather strong results on the spectrum of transfer operators, mostly for rational or polynomial one-dimensional maps with rational weights.

4.B. Differentiable expanding case

When the smoothness assumption on the locally expanding dynamics f and the weight g is weakened from analytic to differentiable (i.e., C^r for some $r \geq 1$) one must abandon the beautiful classical techniques of Grothendieck and work essentially as in the case of subshifts of finite type (i.e., prove bounds "by hand," often using Taylor expansions to approach the transfer operators by finite rank operators). This was first done in the C^∞ case by Tangerman [49] who showed that the corresponding zeta function (2.12) was meromorphic in the whole complex plane. Later, Ruelle [47, 48] studied the case of finite differentiability, proving generalized versions of Theorems 1 and 2 above (also for "mixed" transfer operators, obtained by summing – or integrating – over a family of contractions not necessarily related to a single map). Fried [45] pushed the analysis further, obtaining in particular results on the asymptotics of the distribution of eigenvalues.

We also mention in this subsection the articles of Collet-Isola [44] and Holschneider [46] who obtain exact formulas (as opposed to upper bounds) for transfer operators acting on classes of smooth functions. The introduction of wavelet techniques by Holschneider is novel, and allows to consider a large class of Banach spaces.

2.2.5   The Smooth Hyperbolic Case

Let us consider now a real analytic transformation $f : M \rightarrow M$ of a real analytic manifold and a real analytic weight $g : M \rightarrow \mathbb{C}$, assuming that f satisfies some uniform hyperbolicity condition (such as Anosov). As noted above, if the foliations of f are analytic too, one can reduce to the analytic expanding setting of 4.A, in particular the flat determinant $d_g^\flat(z)$ from the traces (2.14) is an entire function (and the zeta function (2.12), which can be written as a quotient of two finite products of flat determinants for modified weights, is meromorphic in the whole complex plane). Foliations being generically only Hölder, this reduction to the noninvertible expanding case only produces zeta functions meromorphic in a disc (and noncompact transfer operators), just like in the case of subshifts of finite type with Lipschitz weight. It is therefore desirable to introduce transfer operators associated to the "full" invertible hyperbolic system. The price to pay is that one needs to introduce Banach spaces of "functions" (actually distributions) that are smooth along unstable manifolds but rough (i.e., functionals over smooth functions) along stable manifolds (because the inverse map is an expansion along the stable directions). For Axiom A dynamical systems this was done first by Rugh [51, 52] (diffeomorphisms of surfaces or flows on 3-manifolds) who applied Grothendieck's theory to show that the associated transfer operators are nuclear, and that the flat determinants $d_g^\flat(z)$ are entire. This approach was subsequently extended by Fried [50] to arbitrary dimensions. In particular, it follows from [50] that the geodesic flow corresponding to an analytic metric of negative curvature on a compact analytic Riemann manifold extends to a meromorphic function on the whole complex plane. The relation between the zeroes of $d_g^\flat(z)$ and the correlation spectrum has been partially explored [53].

More recently, Kitaev [54] has used a similar approach to study the flat determinants associated to C^r Anosov diffeomorphisms and C^r weights, without assuming additional smoothness of the foliations. (In fact, Kitaev studies mixed transfer operators obtained by summing over maps which preserve the same cone fields.) Here, everything must be done "by hand" and the Banach space of distributions on which the operator acts is again one of the key difficulties. We also mention Liverani's paper [55] on the decay of correlation of certain differentiable hyperbolic systems (using Hilbert-type metrics on Birkhoff cones of functions), although it is not connected to the zeta-function approach, because the philosophy involved in the definition of the Birkhoff cones is very similar to the ideas of Rugh and Kitaev.

2.2.6 The One-dimensional Case

The theory for piecewise monotone interval maps *without* assuming the existence of a finite Markov partition, was developed in parallel to that of Axiom A systems. The natural Banach space preserved by the transfer operator is the space of functions of bounded variation (as was already observed by Lasota and Yorke, see the very nice survey [58] for references). There is no flat trace in general, and one simply uses the counting trace and the "ordinary" zeta function (2.12). Hofbauer and Keller [59, 60] obtained crucial results, using the countable Markov towers of Hofbauer, under certain conditions. A rather general result which is the analogue of Theorem 1 in the Introduction above (assuming that the weight g is of bounded variation and the partition into intervals of monotonicity is generating) was later obtained by Baladi and Keller [57], using again the Hofbauer tower technique. In [66] Ruelle subsequently used another approach (working with Markovian extensions with finitely many symbols, and setting the weight equal to zero to describe the infinite grammar of the original dynamics) and obtained further results (in particular relating the spectral radius to topological entropies). Reference [65] contains in particular the extension to the case where a finite or countable set of branches (not necessarily associated to a single interval map) is considered (see [75] and especially [76], [71] for variants and simplifications of the proof in [65]). The case of one complex dimension is considered in [67] (see also [71]).

The case of an interval map (e.g. with two branches) which is expanding except at a single neutral fixed point ($f(x) = x$ and $f'(x) = 1$) is interesting in view of applications to physics. Prellberg [64] and then Isola [61] show how to study the original map with the help of an auxiliary induced map which is piecewise (uniformly) expanding (with countably many intervals of monotonocity). Isola [61] proves in certain cases that correlation functions associated with the unique absolutely continuous invariant measure decay polynomially for suitable observables.

The case of Collet-Eckmann (or Benedicks Carleson)-type unimodal maps with a critical point (in particular from the logistic family) was studied independently by Keller-Nowicki [62] and Lai-Sang Young [69] who used tower extensions to remove the singularities in the weight $g = 1/|f'|$ of the transfer operator. Ruelle [68] introduced some methods from several complex variables to study directly the zeta function. Pollicott [63] pushed this approach further.

2.2.7 The One-dimensional Case: Kneading Operator Approach

One of the main results in the classical paper [74] by Milnor and Thurston is a formula for the (unweighted) "negative" zeta function of a piecewise monotone interval map

$$\zeta^-(z) = \exp \sum_{n \geq 1} \frac{z^n}{n} 2 \,\#\mathrm{Fix}^- f^n \,, \tag{2.17}$$

where $\mathrm{Fix}^- f^n$ is the number of fixed points of f^n which lie in an interval of monotonicity of f^n where f^n is decreasing (such intersections are transverse and therefore always form a finite set if f has finitely many intervals of monotonicity). This formula simply said that $1/\zeta^-(z)$ is the determinant of a finite matrix (the *kneading matrix*) whose coefficients are power series (the *kneading invariants*) in z associated to the orbits of the turning points.

In [72], [75], [70], and finally [73] this result of Milnor and Thurston was extended by Baladi and Ruelle to the weighted case, where the weight g is supposed to be of bounded variation, and where the branches considered do not necessarily come from a single map. (The transfer operators act on functions of bounded variation.) Since this situation includes the case where infinitely many (even uncountably many) fixed points exist, a new trace must be introduced, which is called the *sharp trace*. Ruelle [76] also considered similar cases where the weight is smoother. We refer to the reviews [9, 10] for more details.

Complex analogues of this approach are presented in [71].

Bibliography

0. Foundations

[1] M. Artin and B. Mazur. On periodic points. *Ann. of Math. (2)*, 21:82–99, 1965.

[2] N. Dunford and J.T. Schwartz. *Linear Operators, Part I.* Wiley Classics Library Edition, New York, 1988.

[3] A. Grothendieck. *Produits tensoriels topologiques et espaces nucléaires (Mem. Amer. Math. Soc. 16).* Amer. Math. Soc., 1955.

[4] A. Grothendieck. La théorie de Fredholm. *Bull. Soc. Math. France*, 84:319–384, 1956.

[5] T. Kato. *Perturbation Theory for Linear Operators.* Second Corrected Printing of the Second Edition. Springer-Verlag, Berlin, 1984.

[6] S. Smale. Differentiable dynamical systems. *Bull. Amer. Math. Soc.*, 73:747–817, 1967.

1. Surveys

[7] V. Baladi. Comment compter avec les fonctions zêta? *Gaz. Math.*, 47:79–96, 1991.

[8] V. Baladi. Dynamical zeta functions. In B. Branner and P. Hjorth, editors, *Real and Complex Dynamical Systems*, pages 1–26. Kluwer Academic Publishers, 1995.

[9] V. Baladi. Periodic orbits and dynamical spectra. In preparation.

[10] V. Baladi. Dynamical zeta functions and generalised Fredholm determinants. *With an appendix written with D. Ruelle,* Some properties of zeta functions associated with maps in one dimension. *XIth International Congress of Mathematical Physics (Paris, 1994)*, pages 249–260. Internat. Press, Cambridge, 1995.

[11] P. Cvitanović et al. Classical and quantum chaos: A cyclist treatise. http://www.nbi.dk/ ~predrag/QCcourse, 1996.

[12] N.E. Hurt. Zeta functions and periodic orbit theory: a review. *Resultate Math.*, 23:55–120, 1993.

[13] D. Ruelle. Functional determinants related to dynamical systems and the thermodynamic formalism (Lezioni Fermiane, Pisa). IHES preprint (1995).

2. Applications

[14] R. Artuso, E. Aurell, and P. Cvitanović. Recycling of strange sets: I. Cycle expansions, II. Applications *Nonlinearity* 3:325–359 and 361–386, 1990.

[15] F. Christiansen, P. Cvitanović, and H.H. Rugh. The spectrum of the period-doubling operator in terms of cycles. *J. Phys. A*, 23:L713–L717, 1990.

[16] Y. Jiang, T. Morita, and D. Sullivan. Expanding direction of the period doubling operator. *Comm. Math. Phys.*, 144:509–520, 1992.

[17] A. Knauf. On a ferromagnetic spin chain. *Comm. Math. Phys.*, 153:77–115, 1993.

[18] E.B. Vul, K.M. Khanin, and Ya.G. Sinai. Feigenbaum universality and the thermodynamic formalism. *Russian Math. Surveys*, 39:1–40, 1984.

3. Subshifts of finite type and Axiom A

[19] R. Bowen. *Equilibrium states and the ergodic theory of Anosov diffeomorphisms*. Springer (Lecture Notes in Math., Vol. **470**), Berlin, 1975.

[20] R. Bowen and O.E. Lanford III. Zeta functions of restrictions of the shift transformation. *Proc. Sympos. Pure Math.*, 14:43–50, 1970.

[21] R. Bowen and D. Ruelle. The ergodic theory of Axiom A flows. *Invent. Math.*, 29:181–202, 1975.

[22] D. Dolgopyat. On decay of correlations in Anosov flows. Preprint, 1996.

[23] D. Dolgopyat. Prevalence of rapid mixing for hyperbolic flows. Preprint, 1996.

[24] G. Gallavotti. Funzioni zeta ed insiemi basilari. *Accad. Lincei Rend. Sc. fis., mat. e nat.*, 61:309–317, 1976.

[25] N.T.A. Haydn. Gibbs functionals on subshifts. *Comm. Math. Phys.*, 134:217–236, 1990.

[26] N.T.A. Haydn. Meromorphic extension of the zeta function for Axiom A flows. *Ergodic Theory Dynamical Systems*, 10:347–360, 1990.

[27] A. Manning. Axiom A diffeomorphisms have rational zeta functions. *Bull. London Math. Soc.*, 3:215–220, 1971.

[28] W. Parry and M. Pollicott. *Zeta functions and the periodic orbit structure of hyperbolic dynamics*. Société Mathématique de France (Astérisque, vol. **187–188**), Paris, 1990.

[29] M. Pollicott. On the rate of mixing of Axiom A flows. *Invent. Math.*, 81:413–426, 1985.

[30] M. Pollicott. Meromorphic extensions of generalised zeta functions. *Invent. Math.*, 85:147–164, 1986.

[31] D. Ruelle. *Thermodynamic Formalism*. Addison Wesley, Reading, MA, 1978.

[32] D. Ruelle. Flots qui ne mélangent pas exponentiellement. *C. R. Acad. Sci. Paris Sér. I Math*, 296:191–193, 1983.

[33] D. Ruelle. Resonances for Axiom A flows. *J. Differential Geom.*, 25:99–116, 1987.

[34] D. Ruelle. One-dimensional Gibbs states and Axiom A diffeomorphisms. *J. Differential Geom.*, 25:117–137, 1987.

4. The smooth expanding case

4.A Analytic expanding case

[35] A. Eremenko, G. Levin, and M. Sodin. On the distribution of zeros of a Ruelle zeta function. *Comm. Math. Phys.*, 159:433–441, 1994.

[36] D. Fried. The zeta functions of Ruelle and Selberg I. *Ann. Sci. École Norm. Sup. (4)*, 19:491–517, 1986.

[37] G. Levin. On Mayer's conjecture and zeros of entire functions. *Ergodic Theory Dynamical Systems*, 14:565–574, 1994.

[38] G. Levin, M. Sodin, and P. Yuditskii. A Ruelle operator for a real Julia set. *Comm. Math. Phys.*, 141:119–131, 1991.

[39] G. Levin, M. Sodin, and P. Yuditskii. Ruelle operators with rational weights for Julia sets. *J. Analyse Math.*, 63:303–331, 1994.

[40] D. Mayer. On a ζ-function related to the continued fraction transformation. *Bull. Soc. Math. France*, 104:195–203, 1976.

[41] D. Mayer. On the thermodynamic formalism for the Gauss map. *Comm. Math. Phys.*, 130:311–333, 1990.

[42] D. Mayer. Continued fractions and related transformations. In T. Bedford, M. Keane, and C. Series, editors, *Ergodic Theory, Symbolic Dynamics and Hyperbolic Spaces*. Oxford University Press, Oxford, 1991.

[43] D. Ruelle. Zeta functions for expanding maps and Anosov flows. *Inv. Math.*, 34:231–242, 1976.

4.B Differentiable expanding case

[44] P. Collet and S. Isola. On the essential spectrum of the transfer operator for expanding Markov maps. *Comm. Math. Phys.*, 139:551–557, 1991.

[45] D. Fried. The flat-trace asymptotics of a uniform system of contractions. *Ergodic Theory Dynamical Systems*, 15:1061–1073, 1995.

[46] M. Holschneider. Wavelet analysis of transfer operators acting on n-dimensional Hölder Besov Zygmund Triebel spaces. Preprint, 1996.

[47] D. Ruelle. The thermodynamic formalism for expanding maps. *Comm. Math. Phys.*, 125:239–262, 1989.

[48] D. Ruelle. An extension of the theory of Fredholm determinants. *Publ. Math. I.H.E.S.*, 72:175–193, 1990.

[49] F. Tangerman. Meromorphic continuation of Ruelle zeta function. (Ph.D. Thesis, Boston University, unpublished), 1986.

5. The smooth hyperbolic case

5.A Hyperbolic analytic case

[50] D. Fried. Meromorphic zeta functions for analytic flows. *Comm. Math. Phys.*, 174:161–190, 1995.

[51] H.H. Rugh. The correlation spectrum for hyperbolic analytic maps. *Nonlinearity*, 5:1237–1263, 1992.

[52] H.H. Rugh. Generalized Fredholm determinants and Selberg zeta functions for Axiom A dynamical systems. *Ergodic Theory Dynamical Systems*, 16:805–819, 1996.

[53] H.H. Rugh. Fredholm determinants for real-analytic hyperbolic diffeomorphisms of surfaces. *XIth International Congress of Mathematical Physics (Paris, 1994)*, pages 297–303. Internat. Press, Cambridge, 1995.

5.B Hyperbolic differentiable case

[54] A. Kitaev. Fredholm determinants for hyperbolic diffeomorphisms of finite smoothness. Preprint, 1995.

[55] C. Liverani. Decay of correlations. *Ann. of Math. (2)*, 142:239–301, 1995.

6. The one–dimensional case

[56] V. Baladi, Y. Jiang, and O.E. Lanford III. Transfer operators acting on Zygmund functions. *Trans. Amer. Math. Soc.*, 348:1599–1615, 1996.

[57] V. Baladi and G. Keller. Zeta functions and transfer operators for piecewise monotone transformations. *Commun. Math. Phys.*, 127:459–479, 1990.

[58] P. Collet. Some ergodic properties of maps of the interval. In R. Bamón, J.-M. Gambaudo, and S. Martinez, editors, *Proceedings, Dynamical Systems and Frustrated Systems – 1991.* to appear.

[59] F. Hofbauer and G. Keller. Ergodic properties of invariant measures for piecewise monotonic transformations. *Math Z.*, 180:119–140, 1982.

[60] F. Hofbauer and G. Keller. Zeta-functions and transfer-operators for piecewise linear transformations. *J. Reine Angew. Math.*, 352:100–113, 1984.

[61] S. Isola. Dynamical zeta functions and correlation functions for intermittent interval maps. Preprint, 1996.

[62] G. Keller and T. Nowicki. Spectral theory, zeta functions and the distribution of periodic points for Collet-Eckmann maps. *Comm. Math. Phys.*, 149:31–69, 1992.

[63] M. Pollicott. One-dimensional maps via complex analysis in several variables. *Israel J. Math*, 91:317–339, 1995.

[64] T. Prellberg. *Maps of the interval with indifferent fixed points: thermodynamic formalism and phase transitions.* PhD thesis, Virginia Polytechnic Institute and State University, 1991.

[65] D. Ruelle. Spectral properties of a class of operators associated with maps in one dimension. *Ergodic Theory Dynamical Systems*, 11:757–767, 1991.

[66] D. Ruelle. *Dynamical Zeta Functions for Piecewise Monotone Maps of the Interval,* CRM Monograph Series, Vol. 4. Amer. Math. Soc., Providence, NJ, 1994.

[67] D. Ruelle. Spectral properties of a class of operators associated with conformal maps in two dimensions. *Comm. Math. Phys.*, 144:537–556, 1992.

[68] D. Ruelle. Analytic completion for dynamical zeta functions. *Helv. Phys. Acta*, 66:181–191, 1993.

[69] L.-S. Young. Decay of correlations for certain quadratic maps. *Comm. Math. Phys.*, 146:123–138, 1992.

7. The one-dimensional case: kneading operator approach

[70] V. Baladi. Infinite kneading matrices and weighted zeta functions of interval maps. *J. Functional Analysis*, 128:226–244, 1995.

[71] V. Baladi, A. Kitaev, D. Ruelle, and S. Semmes. Sharp determinants and kneading operators for holomorphic maps. IHES preprint, 1995.

[72] V. Baladi and D. Ruelle. An extension of the theorem of Milnor and Thurston on the zeta functions of interval maps. *Ergodic Theory Dynamical Systems*, 14:621–632, 1994.

[73] V. Baladi and D. Ruelle. Sharp determinants. *Invent. Math.*, 123:553–574, 1996.

[74] J. Milnor and W. Thurston. Iterated maps of the interval. preprint (Published in Dynamical Systems (Maryland 1986-87), J.C. Alexander, ed. Lecture Notes in Math. Vol. 1342 (1988), Springer-Verlag, Berlin), 1977.

[75] D. Ruelle. Functional equation for dynamical zeta functions of Milnor-Thurston type. *Comm. Math. Phys.*, 175:63–88, 1996.

[76] D. Ruelle. Sharp determinants for smooth interval maps. IHES preprint, 1995.

Chapter 3

Irregular Scattering

> *"Les circonstances que nous venons de rencontrer se retrouveront-elles dans d'autres problèmes de Mécanique? Se présenteront-elles, en particulier, dans l'étude des mouvements des corps célestes? C'est ce qu'on ne saurait affirmer. Il est probable, cependant, que les résultats obtenus dans ces cas difficiles seront analogues aux précédents, au moins par leur complexité."*
>
> J. Hadamard [6]

Scattering experiments are a primary source of our knowledge about elementary particles, atoms and molecules. Similarly celestial bodies are scattered by the sun or the whole solar system.

But whereas two-body scattering (and scattering by two centres which is a restricted three-body system) can be solved analytically, multi-body scattering has dynamical aspects which are much more complicated. Yet it turns out that multi-body scattering is in some cases simpler than the interaction of particles that are confined forever in some bounded region of configuration space.

The basic phenomena, in particular the occurrence of Cantor sets of bounded orbits, were already analysed by Hadamard in the late 19th century. He considered geodesics on a negatively curved non-compact surface with several 'horns' going to infinity.

In the last decade classical and quantum irregular scattering has been the object of extensive study. For the physics-oriented literature we refer to the review articles of Eckhardt [2] (1988), Smilansky [19] (1989), and Tél [20] (1990).

3.1 Notions of Classical Potential Scattering

The scattering of one classical particle by the smooth potential $V : M \to \mathbb{R}$ on a *configuration space* $M = \mathbb{R}_q^n - B$, B bounded, is described by a Hamiltonian function of the form

$$H_{\mathrm{cl}}(\mathbf{p}, \mathbf{q}) := \tfrac{1}{2}|\mathbf{p}|^2 + V(\mathbf{q}), \qquad (\mathbf{p}, \mathbf{q}) \in P := T^*M. \tag{3.1}$$

Typically the configuration space is the whole $\mathbb{R}^n_{\mathbf{q}}$, but in celestial mechanics we would like to treat *Coulombic* potentials of the form

$$V(\mathbf{q}) = \sum_{j=1}^{m} \frac{-Z_j}{|\mathbf{q} - \mathbf{s}_j|}, \tag{3.2}$$

so that in that case $B := \{\mathbf{s}_1, \ldots, \mathbf{s}_m\}$ consists of the centres of attraction. Depending on the context, the constants $Z_j > 0$ are interpreted as charges of nuclei or as masses of celestial bodies. Collision orbits are smoothly regularized by backward scattering.

Another case we will discuss is scattering by *obstacles*, that is, non-overlapping bounded regions $B_1, \ldots, B_m \subset \mathbb{R}^n_{\mathbf{q}}$ with smooth boundaries $\partial \bar{B}_j$. In that case $M = \mathbb{R}^n_{\mathbf{q}} - B$ with $B := \cup_{j=1}^{m} B_j$, and we continue the motion across the phase space boundary $\partial P = \mathbb{R}^n_{\mathbf{p}} \times \partial M$ by the rule 'reflection angle $= -$ incidence angle' (that is, $(\mathbf{p}, \mathbf{q}) \mapsto (\mathbf{p} - 2(\mathbf{n}(\mathbf{q}), \mathbf{p})\mathbf{n}(\mathbf{q}), \mathbf{q})$, where $\mathbf{n}(\mathbf{q})$ is the outward directed unit normal vector at $\mathbf{q} \in \partial M$). If there is no additional potential, then in between these collisions the motion is free, and thus after a rescaling of time by the factor $\sqrt{E}$ the motions for different energies E coincide.

In any case we assume that the potential V decays at infinity. It is called *short-range* if the force $-\nabla V(\mathbf{q}) = \mathcal{O}(|\mathbf{q}|^{-2-\varepsilon})$ for some $\varepsilon > 0$, that is, if it decays faster than the Coulomb potential. (In fact, one should add a kind of Lipschitz condition at infinity of the form

$$|\nabla V(\mathbf{q}_1) - \nabla V(\mathbf{q}_2)| = \mathcal{O}\left(\frac{|\mathbf{q}_1 - \mathbf{q}_2|}{|\mathbf{q}_1|^{2+\varepsilon}}\right),$$

see Simon [16].)

Then we can compare the motion $\Phi^t : P \to P$ generated by H_{cl} with the *free motion*

$$\Phi^t_\infty : P_\infty \to P_\infty, \qquad \Phi^t_\infty(\mathbf{p}, \mathbf{q}) := (\mathbf{p}, \mathbf{q} + t\mathbf{p}) \tag{3.3}$$

generated by the Hamiltonian function $H_\infty(\mathbf{p}, \mathbf{q}) := \tfrac{1}{2}|\mathbf{p}|^2$ on $P_\infty := T^*\mathbb{R}^n_{\mathbf{q}}$, and the pointwise limits

$$\Omega^\pm := \lim_{t \to \pm\infty} \Phi^{-t} \circ \mathrm{Id} \circ \Phi^t_\infty$$

of the *Møller transformations* $\Omega^\pm : \left(\mathbb{R}^n_{\mathbf{p}} - \{\mathbf{0}\}\right) \times \mathbb{R}^n_{\mathbf{q}} \to P$ exist[1].

We denote their inverse by $\Omega^\pm_*$, see Figure 3.1.

Here the two phase spaces P and P_∞ are canonically identified by Id away from the bounded region B.

---

[1]We must exclude the zero momentum subspace from the domain of definition, since there the free particle does not go to infinity.

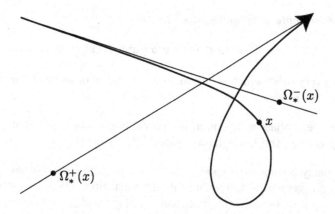

Figure 3.1: The Møller transformation

The Møller transformations preserve Liouville measure. If there are no obstacles ($B = \emptyset$) and under additional conditions on the behaviour of V at infinity, they are smooth canonical transformations.

If we describe a scattering of a charged classical particle by a molecule, we add to the Coulombic potential (3.2) of the nuclei a smooth potential $W : \mathbb{R}_{\mathbf{q}}^n \to \mathbb{R}$ which describes the shielding effect of the bounded electrons. Then the total potential V is assumed to have the asymptotic form

$$V(\mathbf{q}) \sim -Z_\infty/|\mathbf{q}|,$$

$Z_\infty \in \mathbb{R}$ being the net charge of the nucleus. If then $V(\mathbf{q}) + Z_\infty/|\mathbf{q}|$ is of short range, we can define Møller transformations as above by comparing with the 'free' Kepler flow generated by

$$H_\infty(\mathbf{p}, \mathbf{q}) := \tfrac{1}{2}|\mathbf{p}|^2 - Z_\infty/|\mathbf{q}|, \tag{3.4}$$

see Klein and Knauf [9].

The ranges $s^\pm$ of the Møller transformations $\Omega^\pm$ lie on orbits which extend to spatial infinity at times $t \to \pm\infty$.

On the other hand, we may define the set

$$b^\pm := \left\{ (\mathbf{p}_0, \mathbf{q}_0) \in P \,\middle|\, \limsup_{t \to \pm\infty} |\mathbf{q}(t, (\mathbf{p}_0, \mathbf{q}_0))| < \infty \right\}$$

of phase space points which belong to trajectories that are bounded in positive resp. negative time.

The points in $b := b^+ \cap b^-$ are sometimes called *bound states*, the points in $s := s^+ \cap s^-$ *scattering states*.

It turns out that (see Hunziker [7])

- we have *asymptotic completeness* [2] in the sense

$$s^{\pm} = \{x \in P \mid x \notin b^{\pm} \text{ and } H(x) > 0\},$$

that is, every orbit which is unbounded in positive or negative time is asymptotic to free motion.

- Almost every phase space point is either a bound state or a scattering state. More precisely, the Liouville measure $\lambda(P - (b \cup s)) = 0$.

The first property is somewhat specific to the scattering of one particle by a potential and does not generalize to true multibody-scattering. There the potential does not decay along directions in the multibody configuration space where clusters of particles stick together. There Mather and McGehee found examples of messenger orbits with *finite* lim inf but *infinite* lim sup, see [12].

In quantum mechanics the analogous problem of asymptotic completeness has been a major question in mathematical physics and has been solved recently. See Hunziker and Sigal [8] for an overview.

The second property does not exclude the case of orbits which are bounded in the past and scattering in the future or vice versa. We only know that these orbits are exceptional in the measure sense. Nevertheless such orbits influence their phase space neighbourhood in the sense that orbits with a similar future stay bounded for a long time in the past. It turns out – and we shall come back to that point – that the quantal phenomenon of scattering resonances can be traced back to the existence of orbits in the exceptional symmetric difference $s^{+} \triangle s^{-}$.

We may combine the Møller transformations to define the *scattering transformation*

$$S := \Omega_{*}^{+} \circ \Omega^{-} : \Omega_{*}^{-}(s) \to P_{\infty},$$

see Figure 3.1. So the scattering transformation preserves the Liouville measure and is a smooth canonical transformation (again under appropriate conditions on V). We write $\Omega_{*}^{\pm} = (\mathbf{p}^{\pm}, \mathbf{q}^{\pm})$

Now in a typical scattering experiment, one sends in a whole beam of particles whose initial momentum $\mathbf{p}^{-}$ is known, but whose initial impact parameters are (approximately) equidistributed. By *impact parameter* we mean the component $\mathbf{q} - \frac{(\mathbf{q},\mathbf{p})}{(\mathbf{p},\mathbf{p})}\mathbf{p}$ of the position $\mathbf{q}$ perpendicular to the direction of the particle. Note that this quantity is independent of time for the free flow (3.3).

Then the intensity of the outgoing beam with final momentum $\mathbf{p}^{+}$ is measured.

Now by energy conservation $|\mathbf{p}^{-}| = |\mathbf{p}^{+}| = \sqrt{2E}$. So an initial orbit is characterized by its energy E, initial direction $\theta^{-} := \mathbf{p}^{-}/|\mathbf{p}^{-}| \in S^{n-1}$ and its initial impact parameter $\mathbf{l}^{-}$ which can be interpreted as a point on the tangent space $T_{\theta^{-}}S^{n-1}$ of that sphere. The final direction $\theta^{+} := \mathbf{p}^{+}/|\mathbf{p}^{+}| \in S^{n-1}$ is, for given

---

[2] There are other definitions of asymptotic completeness in classical scattering theory.

data E and θ^-, a function θ^+_{E,θ^-} of $\mathbf{l}^-$. The *cross section measure* of a measurable set $\Theta \subset S^{n-1}$ of final directions is then given by

$$\sigma_{E,\theta^-}(\Theta) := \lambda \left(\theta^+_{E,\theta^-} \right)^{-1} (\Theta).$$

The Radon-Nikodym derivative of that measure is called the *differential cross section* (See Reed and Simon [15], Vol. III).

3.2 Centrally Symmetric Potentials

We shortly explain these notions for the simple example of a centrally symmetric potential $V \in C^\infty(\mathbb{R}^n_{\mathbf{q}})$ with $V(\mathbf{q}) \equiv \tilde{V}(|\mathbf{q}|)$. Then the motion takes place in the configuration space plane (or line) spanned by $\mathbf{p}$ and $\mathbf{q}$.

Thus dynamical questions are reduced to the case of $n = 2$ dimension. The differential cross section depends only on the angle between the ingoing and outgoing direction (However, if we express in the usual way the cross section as a function of this angle, then that expression depends explicitly on the dimension $n!$).

Introduction of polar coordinates (r, φ) in the configuration plane with $q_1 = r\cos(\varphi)$, $q_2 = r\sin(\varphi)$ leads to the conjugate momenta $p_r = (\mathbf{q}, \mathbf{p})/|\mathbf{q}|$ and $p_\varphi = p_2 q_1 - p_1 q_2$ which allow to write the Hamiltonian function in the form

$$\tfrac{1}{2} p_r^2 + \tilde{V}(r) + \frac{p_\varphi^2}{2r^2}$$

which is independent of the angle φ. Thus the angular momentum p_φ is a constant of the motion, and we are led to the radial Hamiltonian function

$$H_l(r, p_r) := \tfrac{1}{2} p_r^2 + \tilde{V}_l(r) \tag{3.5}$$

with the *effective potential* $\tilde{V}_l(r) := \tilde{V}(r) + l^2/2r^2$ see Figure 3.2. From the example of a Yukawa potential $\tilde{V}(r) := -\frac{e^{-r}}{r}$ we see that $\tilde{V}_l$ may have positive relative minima even if V is negative. In that case the bound states of positive energy have positive measure.

If we fix the energy E and the angular momentum l of the particle, we have from (3.5) the equation

$$\dot{r} = \pm\sqrt{2(E - \tilde{V}_l(r))}$$

for the radial velocity. On the other hand the angular momentum equals $p_\varphi = r^2 \dot{\varphi} = l$, so that the total scattering angle $\Delta\theta := \theta^+ - \theta^-$ is given by

$$\theta^+ - \theta^- = 2 \int_{r_{\min}}^{\infty} \frac{l/r^2}{\sqrt{2(E - \tilde{V}_l(r))}} dr$$

with $r_{\min}$ the *maximal* solution of the equation $\tilde{V}_l(r) = E$.

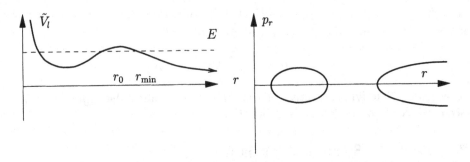

Figure 3.2: The effective potential $\tilde{V}_l$ (left); the energy shell $H_l^{-1}(E)$ (right)

Example. For the Coulomb potential $\tilde{V}(r) = -Z/r$ with charge $Z \in \mathbb{R}$ we obtain

$$\Delta\theta = 2\arccos\left(\frac{\text{sign}(Z)}{\sqrt{1 + 2El^2/Z^2}}\right)$$

so that the absolute value of $\Delta\theta$ is independent of the sign of the charge Z. Note the monotone dependence on the angular momentum l.

In the case of centrally symmetric potentials the differential cross section can be written in the form

$$\frac{d\sigma_{E,\theta-}}{d\theta^+}(\theta^+) = \frac{1}{\sqrt{2E}} \sum_{l:\Delta\theta(l)=\theta^+-\theta^-} \left|\left(\frac{d\Delta\theta}{dl}\right)^{-1}\right|$$

Example. For the Coulomb potential in $n = 2$ dimensions we obtain the (2D.) Rutherford cross section.

$$\frac{d\sigma_{E,\theta-}}{d\theta^+}(\theta^+) = \frac{|Z|}{4E}\frac{1}{\sin^2((\theta^+ - \theta^-)/2)}. \tag{3.6}$$

If the potential V is smooth (and non-zero) then, unlike in the above example of the Coulomb potential, we see that the deflection angle is not monotone in the angular momentum, since for energies $E > V_{\max}$ there is no deflection ($\Delta\theta = 0$) both for $l = 0$ and as $|l| \to \infty$. So there exist extrema of the function $l \mapsto \Delta\theta(l)$.

These extrema show up in the differential cross section which exhibit singularities at the corresponding angles. A similar phenomenon for the scattering of light by rain drops is responsible for the rainbow.

Geometrically one may interpret these singularities as singularities of the configuration space projection of the Lagrange manifold $\Omega^-(\{\mathbf{p}^-\} \times \mathbb{R}_q^n) \subset P$ consisting of the orbits with a given initial angular momentum $\mathbf{p}^-$ and arbitrary initial impact parameter.

There exists a well-developed theory of the semiclassical quantum mechanics of these fold singularities, see Duistermaat [4].

Centrally symmetric potentials may also lead to captured orbits which are scattering in the past and bounded in the future. Namely for a given value l of the angular momentum the effective potential $\tilde{V}_l$ may have a relative maximum at r_0, see Figure 3.2. If the energy E equals the value $\tilde{V}_l(r_0)$, then the orbits with these energies and angular momenta are positive resp. negative asymptotic to the unstable circular orbit with radius r_0. In that case the symmetric difference $s^+ \Delta s^- \neq \emptyset$.

3.3 Scattering by Convex Obstacles

> *"... Each pearl by imaging those immediately adjacent to it images*
> *the infinity of pearls in the outer space of the whole net,*
> *for each pearl is the bearer of its neighbour's image."*
> *From the Avatamsaka Sutra, cited after [1]*

We consider now in some detail the situation where, instead of being deflected by a potential, the particle scatters at strictly convex obstacles $B_1, \ldots, B_m \subset \mathbb{R}^n_{\mathbf{q}}$, see Figure 3.3.

For $m \geq 3$ obstacles this leads to the simplest class of mechanical systems which exhibit irregular scattering. It is also similar to the well-known example of an ergodic system, the Sinai Billiard [17]. However, due to the unboundedness of configuration space, proofs are much simpler in the scattering problem.

Between the collisions the motion in configuration space has constant velocity, and we normalize the speed $|\mathbf{p}| := 1$, that is, $E := \frac{1}{2}$. Thus the energy shell $\Sigma := H_{\text{cl}}^{-1}(\frac{1}{2})$ is homeomorphic to $S^{n-1} \times M/ \sim$, with identification $(\mathbf{p}, \mathbf{q}) \sim (\mathbf{p} - 2(\mathbf{n}(\mathbf{q}), \mathbf{p})\mathbf{n}(\mathbf{q}), \mathbf{q})$ of points $\mathbf{q} \in \partial M$ at the boundary. For simplicity we consider the situation where no obstacles *shadow* each other, that is, every straight line in the plane intersects at most two obstacles $\bar{B}_j$.

As the above citation suggests, we may think of the mechanical system not only as a special kind of *billiard*, but also in terms of *geometric optics* with mirrors $\partial \bar{B}_j$. The language of optics has the advantage that not only the individual orbit (light ray) but families of such rays are taken into consideration. These are then (de)focused by the mirrors.

For general dimension n we use the cotangent bundle $T^* \partial M$ of the boundary ∂M of the obstacles to describe the reflection data. This is possible since the component of the momentum tangential to ∂M is not altered during the reflection. $T^* \partial M$ is a $2(n-1)$-dimensional submanifold of $T^* \mathbb{R}^n_{\mathbf{q}}$ and has its own natural symplectic two-form ω. We fixed the energy to be $E = \frac{1}{2}$ so that $|\mathbf{p}| = 1$. This implies that the tangential component $\mathbf{p} - (\mathbf{n}(\mathbf{q}), \mathbf{p})\mathbf{n}(\mathbf{q})$ of the momentum $\mathbf{p}$ is contained in the unit disk of $T^*_{\mathbf{q}} \partial M$. So our *reduced phase space* $\mathcal{P} \subset \Sigma$ is the disk bundle of ∂M, and the Poincaré return map F leaves the symplectic two-form ω invariant.

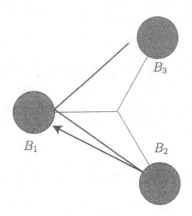

Figure 3.3: Scattering by three disks

More specifically, we can coordinatize for $n = 2$ the curves $\partial \bar{B}_j$ by arc length l. Then the tangential component of the momentum equals $u := \sin \varphi$, and $\omega = \cos \varphi \, dl \wedge d\varphi = dl \wedge du$, $\varphi \in [-\pi/2, \pi/2]$ being the angle between the ray and the outward normal direction of the mirror at p.

The reduced phase space $\mathcal{P} \subset \Sigma$ thus has the form of m disjoint copies of cylinders $\mathcal{P}_j = \{(u, l) \in [-1, 1] \times S^1_j\}$, $S^1_j \cong \partial \bar{B}_j$ being the circle parametrized by arc length l.

We now restrict ourselves to planar scattering ($n = 2$), and its linearization, using *Jacobi coordinates* $(\delta p, \delta q)$ near a flow line (a geodesic in our Euclidean metric) $t \mapsto (\dot{\mathbf{l}}_0(t), \mathbf{l}_0(t)) \in \Sigma$ of the form

$$\mathbf{l}_0(t) := \mathbf{q}_0 + \mathbf{p}_0 t, \quad (\mathbf{p}_0, \mathbf{q}_0) \in \Sigma.$$

These Jacobi coordinates give rise to one-parameter families $\{\mathbf{l}_\varepsilon\}$ of nearby straight lines of the form

$$\mathbf{l}_\varepsilon(t) := (\mathbf{q}_0 + \varepsilon \mathbf{p}_0^\perp \delta q_0) + (\mathbf{p}_0 + \varepsilon \mathbf{p}_0^\perp \delta p_0)t,$$

with the unit vector $\mathbf{p}_0^\perp := \left(\begin{smallmatrix} 0 & 1 \\ -1 & 0 \end{smallmatrix} \right) \mathbf{p}_0$ orthogonal to $\mathbf{p}_0$. Since we are not interested in variations in the direction of the flow, we do not consider variations of $\mathbf{q}_0$ parallel to $\mathbf{p}_0$.

The Jacobi coordinates thus evolve according to

$$\begin{pmatrix} \delta p(t) \\ \delta q(t) \end{pmatrix} = \begin{pmatrix} 1 & 0 \\ t & 1 \end{pmatrix} \begin{pmatrix} \delta p(0) \\ \delta q(0) \end{pmatrix}. \tag{3.7}$$

Because of the normalization $|\mathbf{p}| = 1$ of the speed the variation δp can be interpreted as the variation of the *direction* of the velocity vector $\mathbf{p}$.

It is useful to introduce the *curvature* $k(t) := \delta p(t)/\delta q(t)$ of the family of lines. Through each point $l_0(t)$ of the line passes a curve which is orthogonal to the family l_ε of lines, see Figure 3.4. $k(t)$ is the curvature of that curve at $l_0(t)$. So $k(t) > 0$ (< 0) if the family is *dispersing (focusing)* at time t.

From (3.7) we see that

$$k(t) = \frac{k(0)}{1 + tk(0)}.$$

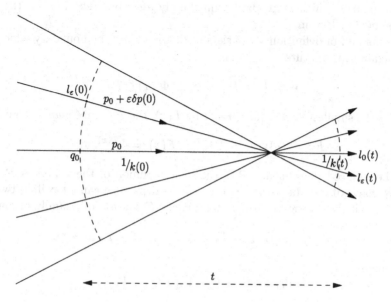

Figure 3.4: Curvature of a family of straight lines

We also want to switch from the variation $(\delta u, \delta l)$ of the boundary coordinates to the Jacobi coordinates $(\delta p^\pm, \delta q^\pm)$ immediately before, respectively after, reflection. This linear symplectic transformation is of the form

$$\begin{pmatrix} \delta p^\pm \\ \delta q^\pm \end{pmatrix} = \begin{pmatrix} \pm 1/\cos\varphi & K(l) \\ 0 & \pm\cos\varphi \end{pmatrix} \begin{pmatrix} \delta u \\ \delta l \end{pmatrix} \tag{3.8}$$

with the curvature K of the boundary ∂M, since $\delta u = \cos(\varphi)\delta\varphi$. Thus during the reflection the Jacobi coordinates transform according to

$$\begin{pmatrix} \delta p^+ \\ \delta q^+ \end{pmatrix} = -\begin{pmatrix} 1 & 2K(l)/\cos\varphi \\ 0 & 1 \end{pmatrix} \begin{pmatrix} \delta p^- \\ \delta q^- \end{pmatrix}. \tag{3.9}$$

If $k^\pm := \delta p^\pm/\delta q^\pm$ denotes the curvature of our family of lines before resp. after reflection, we obtain

$$k^+(t) = k^-(t) + 2K(l)/\cos(\varphi). \tag{3.10}$$

This is nothing but the image equation of geometrical optics, since the focusing length of the mirror at l equals $\cos(\varphi)/2K(l)$.

Now if – as we assumed – the obstacles are strictly convex, the curvature function $K : \partial M \to \mathbb{R}$ is strictly positive and thus bounded by

$$0 < K_{\min} \leq K \leq K_{\max}. \tag{3.11}$$

So by (3.10) a dispersing family remains dispersing after a reflection.

Since the motion is simple between reflections, it is useful to consider a Poincaré map F which maps data immediately after one reflection on the data after the next reflection.

Its domain of definition V consists of the points on the boundary which will collide again in the future:

$$V := \{x \in \mathcal{P} \mid \exists t > 0 : \Phi^t(x) \in \mathcal{P}\}.$$

Denoting the smallest such *return time* t by $T(x)$, the *Poincaré map* F is given by

$$F : V \to W := F(V), \qquad F(x) := \Phi^{T(x)}(x).$$

Since the reduced phase space $\mathcal{P}$ is the disjoint union of the cylinders $\mathcal{P}_j$, and since by convexity of the obstacles the scattered particle cannot collide twice in succession with the same boundary component, F consists of a family of maps

$$F_{j,k} : V_{j,k} \to W_{j,k} := F_{j,k}(V_{j,k}), \qquad (1 \leq j \neq k \leq m) \tag{3.12}$$

with domains

$$V_{j,k} := \{x \in \mathcal{P}_j \cap V \mid F(x) \in \mathcal{P}_k\}.$$

The *time involution* $\mathcal{I} : \mathcal{P} \to \mathcal{P}$, $(\mathbf{p}, \mathbf{q}) \mapsto (-\mathbf{p}, \mathbf{q})$ interchanges domains and images:

$$\mathcal{I}(V_{j,k}) = W_{k,j}, \qquad \mathcal{I} \circ F_{j,k} \circ \mathcal{I} = F_{k,j}^{-1}, \tag{3.13}$$

since the flow Φ^t is reversible.

In the following lemma we show that these domains $V_{j,k}$ (and thus also the images $W_{k,j}$) are deformed quadrangles joining the two boundaries

$$\partial_j^\pm := \{\pm 1\} \times S_j^1 \tag{3.14}$$

of the cylinders $\mathcal{P}_j$, see Figure 3.5:

Lemma 3.1 *There exist strictly decreasing smooth functions*

$$D_{j,k}^- < D_{j,k}^+ : [-1, 1] \to S_j^1$$

with

$$V_{j,k} = \left\{(u, l) \in \mathcal{P}_j \mid D_{j,k}^-(u) \leq l \leq D_{j,k}^+(u)\right\}.$$

Proof. Consider all the rays emanating from the boundary component $\partial\bar{B}_j$ at a given angle φ. This family of rays covers the rest $\mathbb{R}_q^2 - B_j$ of the plane once, without intersections of rays, since B_j was assumed to be convex. This family of rays is parametrized by the arc length l of the point on $\partial\bar{B}_j$. Thus the rays intersecting B_k correspond to a closed interval $[D_{j,k}^-(u), D_{j,k}^+(u)]$ of l-values, with $u = \sin\varphi$. The parameter values $D_{j,k}^\pm(u)$ correspond to rays tangent to $\partial\bar{B}_k$ (here our non-shadowing assumption is used). We can calculate the slopes $\delta l_j/\delta u_j$ of the graphs of $D_{j,k}^\pm$ at (u_j, l_j) by noticing that the Poincaré map $F_{j,k}$ maps the tangent vectors $\binom{\delta u_j}{\delta l_j}$ of these curves to tangent vectors $\binom{\delta u_k}{\delta l_k}$ with $\delta u_k = 0$. Now

$$\begin{pmatrix} \delta u_j \\ \delta l_j \end{pmatrix} = \begin{pmatrix} \cos\varphi_j & -K(l_j) \\ 0 & 1/\cos\varphi_j \end{pmatrix} \begin{pmatrix} \delta p_j^+ \\ \delta q_j^+ \end{pmatrix} \tag{3.15}$$

and

$$\begin{pmatrix} \delta p_j^+ \\ \delta q_j^+ \end{pmatrix} = \begin{pmatrix} 1 & 0 \\ -T(u_j, l_j) & 1 \end{pmatrix} \begin{pmatrix} \delta p_k^- \\ \delta q_k^- \end{pmatrix}$$

with $\binom{\delta p_k^-}{\delta q_k^-} = -K(l_k) \cdot \binom{\delta l_k}{0}$, so that

$$\frac{\delta l_j}{\delta u_j} = -1 \Big/ \left(\cos\varphi_j \left(\frac{\cos\varphi_j}{T(u_j, l_j)} + K(l_j) \right) \right) < 0. \qquad \square$$

Example. We consider three disks B_1, B_2 and B_3 of radius 1 and mutual distance $R > 2$ arranged on an equilaterial triangle. If $R > 4/\sqrt{3} \approx 2.31$, then the disks are not shadowing each other. Their length parameter $l \in [-\pi, \pi]$ is measured relative to the direction to the centre of the triangle, see Figure 3.3. By elementary trigonometry one obtains that $V_{1,2}$ resp. $V_{1,3}$ are bounded by the graphs of the functions

$$D_{1,2}^\pm = D^\pm - \pi/6 \quad \text{resp.} \quad D_{1,3}^\pm = D^\pm + \pi/6$$

with

$$D^\pm(u) := \arcsin\left(\frac{u \pm 1}{R} \right) - \arcsin(u).$$

The other domains and images are symmetry-related, see Figure 3.5.

In the above example, the intersections $W_{i,j} \cap V_{j,k}$ are deformed squares, since the boundary curves $u \mapsto D_{j,k}^\pm(u)$ of $V_{j,k}$ and the boundary curves $u \mapsto D_{j,i}^\pm(-u)$ of $W_{j,i}$ intersect pairwise exactly once.

This is a general property of convex non-shadowing obstacles. Namely,

- by the monotonicity property of $D_{i,j}^\pm$ stated in Lemma 3.1 there can be at most one intersection.

- By the non-shadowing property there is a $u_0 < 1$ such that all points $(u, l) \in W_{j,k} \cap V_{k,l}$ belonging to an orbit which comes from obstacle j and goes to obstacle l have

$$|u| \le u_0.$$

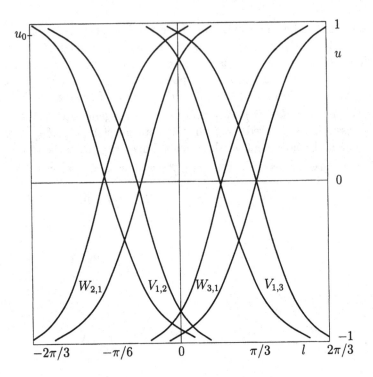

Figure 3.5: The Poincaré surface $\mathcal{P}_1$ for the three-disk system

In other words, as in Fig. 3.5 the sets $W_{i,j} \cap V_{j,k}$ have a positive distance from $\partial_k^{\pm}$.

- $W_{i,j} \cap V_{j,k}$ is non-empty. This is again a consequence of the non-shadowing property. (For example, one can construct such an orbit going from ∂B_i, visiting ∂B_j and then ∂B_k by taking the shortest of the broken line segments connecting points on these three boundaries. This broken segment does not leave M and its reflection angle at ∂B_j is its negative incidence angle. Thus it belongs to a physical trajectory).

Now we iterate the Poincaré map F on the domains

$$V_{f_1,\ldots,f_r} := \left\{ x \in V_{f_1,\ldots,f_{r-1}} \mid F^{r-2}(x) \in V_{f_{r-1},f_r} \right\}.$$

Clearly the images

$$W_{f_1,\ldots,f_r} := F^{r-1}(V_{f_1,\ldots,f_r}) \subset W_{f_2,\ldots,f_r} \subset \mathcal{P}_{f_r}$$

satisfy the relation

$$W_{f_1,\ldots,f_r} = \mathcal{I}(V_{f_r,\ldots,f_1}),$$

generalizing (3.13). The domain $V_{f_1,\ldots,f_r}$ corresponds to the points on the obstacle f_1 which will visit the obstacles $f_2,\ldots,f_r$ in succession. So it is empty unless we have an *admissible sequence* of symbols $f_j \in S := \{1,\ldots,m\}$ with $f_i \neq f_{i+1}$ for $i = 1,\ldots,r-1$.

By an argument similar to the one used in Lemma 3.1, there exist strictly decreasing smooth functions

$$D^-_{f_1,\ldots,f_r} < D^+_{f_1,\ldots,f_r} : [-1,1] \to S^1_{f_1}$$

with

$$V_{f_1,\ldots,f_r} = \left\{ (u,l) \in P_{f_1} \mid D^-_{f_1,\ldots,f_r}(u) \leq l \leq D^+_{f_1,\ldots,f_r}(u) \right\}.$$

Again one considers a family of rays emanating from the f_1st obstacle with constant u, see Figure 3.6.

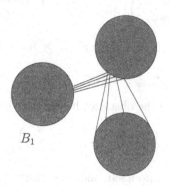

Figure 3.6: A family of rays with constant initial angle emanating from B_1

Lemma 3.2 *The widths of the quadrangles $V_{f_1,\ldots,f_r}$ are exponentially small in r, that is, there exist r-independent constants $0 < c_l < c_u$ and $0 < j_u < j_l$ with*

$$c_l \exp(-j_l \cdot r) \leq D^+_{f_1,\ldots,f_r}(u) - D^-_{f_1,\ldots,f_r}(u) \leq c_u \exp(-j_u \cdot r) \qquad (3.16)$$

for all $u \in [-u_0, u_0]$ (with u_0 defined on page 31) and all admissible sequences $f_1,\ldots,f_r$.

There exist r-independent constants $0 < s_l < s_u$ which bound the slopes of the boundary curves:

$$-s_u \leq \frac{d}{du}D^{\pm}_{f_1,\ldots,f_r}(u) \leq -s_l < 0 \qquad (u \in [-u_0, u_0]). \qquad (3.17)$$

Proof. To derive estimate (3.16), we consider the linearization $D(F^r)$ of the r-fold iterated Poincaré map, which by the chain rule equals

$$D(F^r)(x) = DF(F^{r-1}(x)) \cdot \ldots \cdot DF(x).$$

So we have to multiply r matrices of determinant one which transform the tangent vectors $\binom{\delta u_{f_i}}{\delta l_{f_i}}$ at $F^{i-1}(x)$ to the tangent vectors $\binom{\delta u_{f_{i+1}}}{\delta l_{f_{i+1}}}$.

Instead we will estimate the product $M_r \cdot \ldots \cdot M_1$ of matrices

$$M_i := -\begin{pmatrix} 1 & 2K_{i+1}/\cos\varphi_{i+1} \\ 0 & 1 \end{pmatrix} \cdot \begin{pmatrix} 1 & 0 \\ T_i & 1 \end{pmatrix}$$

which transform the Jacobi coordinates $\binom{\delta p_{f_i}^+}{\delta q_{f_i}^+}$ into the Jacobi coordinates $\binom{\delta p_{f_{i+1}}^+}{\delta q_{f_{i+1}}^+}$.

Here K_i, φ_i and T_i denote the curvature, angle and return time at the ith point $F^{i-1}(x) \in \mathcal{P}_{f_i}$ of collision with an obstacle.

So

$$D(F^r)(x) = \begin{pmatrix} \cos\varphi_r & -K_r \\ 0 & 1/\cos\varphi_r \end{pmatrix} \cdot M_r \cdot \ldots \cdot M_1 \cdot \begin{pmatrix} 1/\cos\varphi_1 & K_1 \\ 0 & \cos\varphi_1 \end{pmatrix}. \quad (3.18)$$

By (3.11) the curvature is bounded, $K_{\max} \geq K \geq K_{\min} > 0$, and similar bounds

$$T_{\max} \geq T \geq T_{\min} > 0 \quad (3.19)$$

exist for the return time, since the obstacles have positive distances.

The cone

$$\mathcal{C} := \left\{ \begin{pmatrix} v_1 \\ v_2 \end{pmatrix} \in \mathbb{R}^2 \,\middle|\, v_1 \cdot v_2 \geq 0 \right\}$$

is mapped by the matrices M_i into itself, and its vectors are stretched: There exist $0 < j_u \leq j_l$ with

$$e^{j_u} \leq \frac{\|M_i\mathbf{v}\|}{\|\mathbf{v}\|} \leq e^{j_l} \quad (\mathbf{v} \in \mathcal{C} \setminus \{0\}).$$

The constants only depend on the bounds (3.11), (3.19) and u_0. The first and the last matrix in (3.18) can change the stretching factor only by a constant c, so that

$$c \cdot e^{j_u r} \leq \frac{\|D(F^r)(x)\mathbf{v}\|}{\|\mathbf{v}\|} \leq c^{-1} \cdot e^{j_l r} \quad (3.20)$$

Each vertical segment of the form

$$\{u\} \times [D_{f_1,\ldots,f_r}^-(u), D_{f_1,\ldots,f_r}^+(u)] \subset V_{f_1,\ldots,f_r}$$

is mapped by F^r on an increasing curve in $W_{f_1,\ldots,f_r} \subset \mathcal{P}_{f_r}$ connecting $\partial_{f_r}^-$ with $\partial_{f_r}^+$, see (3.14). These two circles have distance 2, so that the length of the image curve is ≥ 2. On the other hand, 2 plus the circumference of the closed curve ∂B_{f_r} is an upper bound for that length.

Thus by using (3.20) we see that the length $D_{f_1,\ldots,f_r}^+(u) - D_{f_1,\ldots,f_r}^-(u)$ of our initial segment satisfies a bound of the form (3.16).

The bound (3.17) on the slopes of the boundary curves follows from a geometrical consideration. These curves belong to rays which, after colliding with $\partial \bar{B}_{f_1}, \ldots, \partial \bar{B}_{f_{r-1}}$, meet $\partial \bar{B}_{f_r}$ tangentially. So this family of rays must be focussing. In particular the initial curvature $k_{f_1}^+ = \delta p_{f_1}^+ / \delta q_{f_1}^+$ of the family of rays must be negative, which by (3.15) and (3.11) implies that the slope

$$\frac{\delta l_{f_1}^+}{\delta u_{f_1}^+} = \frac{1}{\cos \varphi_1 (\cos \varphi_1 k_{f_1}^+ - K_1)} \geq -\frac{1}{c_0 K_{\min}}.$$

On the other hand, we must have the inequality

$$\frac{\delta l_{f_1}^+}{\delta u_{f_1}^+} \leq -\frac{1}{(1/T_{\min}) + K_{\max}}$$

since the family of rays has a focal point before meeting $\partial \bar{B}_{f_2}$ if its curvature $k_{f_1}^+ < -1/T_1$. □

Now we describe our scattering system using symbolic dynamics.

Definition 3.3 *A (finite, half-infinite or bi-infinite) sequence $f : A \to \mathcal{S}$ $A \subset \mathbb{Z}$ with the space of symbols $\mathcal{S} = \{1, \ldots, m\}$ is called* admissible *if $f_{i+1} \neq f_i$ for $i, i+1 \in A$.*

For an admissible sequence $f : \mathbb{N}_0 \to \mathcal{S}$ we set

$$V_f := \bigcap_{r \in \mathbb{N}} V_{f_0, \ldots, f_r}.$$

Similarly for an admissible sequence $g : \mathbb{Z} - \mathbb{N} \to \mathcal{S}$

$$W_g := \bigcap_{r \in \mathbb{N}} W_{g_{-r}, \ldots, g_0}.$$

Lemma 3.4 *$V_f \subset \mathcal{P}_{f_0}$ is the graph of a strictly monotone decreasing Lipschitz continuous function*

$$D_f : [-1, 1] \to S^1$$

with the Lipschitz bounds s_l, s_u from Lemma 3.2.

Similarly $W_g \subset \mathcal{P}_{g_0}$ is the graph of a strictly monotone increasing Lipschitz continuous function $D_g : [-1, 1] \to S^1$.

If $f_0 = g_0$, then these two curves intersect in exacly one point

$$V_f \cap W_g$$

(otherwise the two curves lie on different Poincaré surfaces).

Proof. By Definition 3.3 V_f consists of those points $(u, l) \in \mathcal{P}_{f_0}$ which are in all the strips $V_{f_0,\ldots,f_r}$, that is, satisfy all the inequalities

$$D^-_{f_1,\ldots,f_r}(u) \leq l \leq D^+_{f_1,\ldots,f_r}(u), \qquad (r \in \mathbb{N}).$$

But since these strips are nested ($V_{f_0,\ldots,f_r} \supset V_{f_0,\ldots,f_{r+1}}$), and since by estimate (3.16) upper and lower bounds converge uniformly, they have a common limit function D_f. The estimates (3.17) for the slopes of the functions $D^\pm_{f_1,\ldots,f_r}$ are r-independent, so that they are also valid for D_f.

The same argument applies to W_g.

Since D_f is strictly decreasing and D_g is strictly increasing, their graphs intersect at most once. Again, the non-shadowing property implies the existence of an intersection.   □

Now precisely on the sets V_f the iterated Poincaré map F^r is defined for all non-negative times $r \in \mathbb{N}_0$, and similarly $F^r(x)$ is well-defined for all $r \leq 0$ iff x is a point on some curve W_g. But an orbit of the flow Φ^t on the energy shell Σ is bounded for all times precisely if it touches the obstacles again and again. So the set

$$\Lambda := \left\{ x \in \mathcal{P} \,\middle|\, \limsup_{t \in \mathbb{R}} |\Phi^t(x)| < \infty \right\}$$

of points of bounded Φ^t-orbits on the Poincaré surface has the form

$$\Lambda = \bigcup_{f,g \text{ adm.}} (V_f \cap W_g).$$

Our next step will be to analyze this set using symbolic dynamics.

3.4 Symbolic Dynamics

Here we formalize the idea that we can describe a bounded orbit by the succession of its collisions with the obstacles.

The subspace

$$X := \{ f : \mathbb{Z} \to \mathcal{S} \text{ admissible } \} \subset \mathcal{S}^{\mathbb{Z}}$$

of admissible sequences is invariant under the *shift*

$$\sigma : X \to X, \quad \sigma(f)_i := f_{i+1}.$$

We equip the space $\mathcal{S}$ of symbols with the discrete topology, $\mathcal{S}^{\mathbb{Z}}$ with the product topology and X with the subspace topology. By Tychonov's theorem $\mathcal{S}^{\mathbb{Z}}$, being a product of compact spaces, is compact. The product topology, being the coarsest topology on $\mathcal{S}^{\mathbb{Z}}$ for which all the projections $f \mapsto f_i$ are continuous, has a basis of 'open rectangles' $R_{g_i,\ldots,g_j}$, $g_i, \ldots, g_j \in \mathcal{S}$, of the form

$$R_{g_i,\ldots,g_j} := \{ f \in \mathcal{S}^{\mathbb{Z}} \mid f_k = g_k, k = i, \ldots, j \}.$$

In the metric d on $\mathcal{S}^{\mathbb{Z}}$,

$$d(f,g) := \sum_{i \in \mathbb{Z}} 2^{-|i|} \cdot (1 - \delta_{f_i, g_i})$$

an open ball $U_\varepsilon(g) := \{f \in \mathcal{S}^{\mathbb{Z}} \mid d(f,g) < \varepsilon\}$ of radius $\varepsilon := 2^{-n}$ has the property

$$R_{g_{-n-2}, \dots, g_{n+2}} \subset U_\varepsilon(g) \subset R_{g_{-n}, \dots, g_n}$$

and thus generates the product topology.

The shift map σ is continuous, since it maps the open rectangles onto open rectangles.

We describe the points in Λ by their symbol sequence using the projections

$$\Pi^+ : X \to \mathcal{S}^{\mathbb{N}_0}, \quad \text{and} \quad \Pi^- : X \to \mathcal{S}^{\mathbb{Z}-\mathbb{N}}$$

of the bi-infinite sequences to their future resp. past. Then with

$$\mathcal{H} : X \to \Lambda, \qquad f \mapsto V_{\Pi^+(f)} \cap W_{\Pi^-(f)}$$

we will describe Λ using symbolic dynamics[3]. This will enable us to prove statements on the physical dynamics using the simple shift map.

Proposition 3.5 $\mathcal{H} : X \to \Lambda$ *is a Hölder continuous homeomorphism, that is, there exist* $C > 0$, $\alpha > 0$ *with*

$$\operatorname{dist}(\mathcal{H}(f), \mathcal{H}(g)) \leq C(d(f,g))^\alpha, \qquad (f, g \in X). \qquad (3.21)$$

$\mathcal{H}$ *conjugates the dynamics on* Λ *generated by the Poincaré map and the shift dynamics on* X:

$$F \circ \mathcal{H} = \mathcal{H} \circ \sigma.$$

Proof. We use as a distance function dist on the components $\mathcal{P}_j$ the Euclidean distance w.r.t. the (u, l)-coordinates (and define dist so that the distance of points on different components is bounded).

We already argued that $\mathcal{H}$ is onto. $\mathcal{H}$ is one-to-one since for $f, g \in X$ with $f \neq g$ there is an $i \in \mathbb{Z}$ with $f_i \neq g_i$. But then the strips $V_{f_0, \dots, f_i} \ni \mathcal{H}(f)$ and $V_{g_0, \dots, g_i} \ni \mathcal{H}(g)$ are disjoint, and a similar argument applies if $i \leq 0$.

To show Hölder continuity, we can assume $f \neq g$. Let

$$r := \min\{m \in \mathbb{N}_0 \mid f(m) \neq g(m) \text{ or } f(-m) \neq g(-m)\} - 1.$$

So $d(f,g) \geq 2^{-(r+1)}$. Without loss of generality we may assume $r \geq 1$, because otherwise we can satisfy (3.21) by taking a large C.

Then both $\mathcal{H}(f)$ and $\mathcal{H}(g)$ are contained in the 'rectangle'

$$W_{f_{-r}, \dots, f_0} \cap V_{f_0, \dots, f_r}.$$

---

[3]In this definition we identify the one-point sets with the point they contain.

We note that for a suitable r-independent $\tilde{C} > 0$ the diameter of that 'rectangle' is bounded above by

$$\tilde{C} \cdot \exp(-j_u r).$$

Namely by estimate (3.16) any two points (u_1, l_1) and (u_2, l_2) of that set satisfy the estimates $|l_1 - l_2| \le 2c_u \exp(-j_u \cdot r)$, and $|u_1 - u_2| \le 2c_u \exp(-j_u \cdot r)/s_l$.

Thus $\mathcal{H}$ is Hölder continuous with exponent $\alpha = j_u/\ln 2$. $\mathcal{H}$ is a homeomorphism, since X is compact[4].

By induction in r one shows that

$$F(V_{f_0,\dots,f_r}) = V_{f_1,\dots,f_r} \cap W_{f_0,f_1} \text{ and } F(W_{f_{-r},\dots,f_0} \cap V_{f_0,f_1}) = W_{f_{-r},\dots,f_1}.$$

So F shifts the symbol sequence f to the left. $\square$

Figure 3.7: The first steps in the construction of the invariant set Λ

Now we can prove statements about the flow Φ^t by proving corresponding statements about the shift map.

Theorem 3.6   *1. For $m = 1$ obstacle there is no bounded orbit in the energy shell Σ, for $m = 2$ there is exactly one, and for $m \ge 3$ there are uncountably many bounded orbits.*

2. For $m \ge 3$ there are countably many closed orbits, and the number $N_F(T)$ of prime periodic orbits of the Poincaré map F with period smaller than T is asymptotic to

$$N_F(T) \sim \frac{e^{T \ln(m-1)}}{(m-2) \cdot T}.$$

(See Fig. 3.8).

---

[4]A continuous bijection from a compact space to a Hausdorff space is a homeomorphism.

3. *For $m \geq 3$ the invariant set Λ is a Cantor set of measure zero[5].*

4. *There is an orbit which is dense in Λ.*

5. *For a suitable probability measure on Λ the Poincaré map F is mixing if $m \geq 3$.*

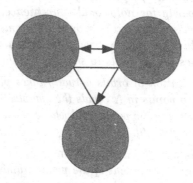

Figure 3.8: The shortest closed orbits for three disks.

Exercise. Prove these statements, using symbolic dynamics. *Hints.*

1. Assuming X to be countable ($X = \{f^1, f^2, \ldots\}$), use for $m \geq 3$ a diagonal argument, i.e. construct $f \in X$ with $f_n \neq f_n^n$, $n \in \mathbb{N}$. Since X is uncountable the orbits are uncountable, too.

2. Use the traces $\operatorname{tr}(A^k)$ for the $m \times m$ *transition matrix A* with $A_{i,k} = 1 - \delta_{i,k}$ to count periodic orbits.

3. To show that X is totally disconnected, by definition one must prove that the *connected component $K(f)$* of $f \in X$ (the union of connected subsets of X containing f) equals $\{f\}$. Use the open rectangles $R_{f_{-n}, \ldots, f_n}$ and their open complements.

   To show that $f \in X$ is non-isolated, construct a sequence of $g^n \in X - \{f\}$ with $g^n \in R_{f_{-n}, \ldots, f_n}$.

4. Construct a $f \in X$ containing all finite admissible subsequences.

5. The two-sided Markov shift measure with stochastic matrix $\frac{A}{m-1}$ (see, e.g. Walters [22]) has this property.

---

[5] *A* Cantor set is a totally disconnected non-void compact set without isolated points. Such sets are homeomorphic to *the* Cantor set.

Remarks 3.7   *1. For the system of three disks one can almost see the Cantor structure of Λ. Instead of the above symbolic description we can equivalently note whether in the next iteration of the Poincaré map we hit the disk left or right of the centre. All sequences of 'left' and 'right' occur. Such a half-infinite sequence can be mapped on the Cantor one-third set by choosing the corresponding subinterval which remains after deletion of the middle interval.*

2. Most statements describe properties of the Poincaré map F instead of the flow Φ^t. This is partly for notational convenience, but there are also real problems if one wants to prove corresponding statements for Φ^t. In particular the mixing property 5 can only be true if the return time T varies on Λ. The same problem arises if one wants to show the analog of property 2 for Φ^t.

Such a desynchronization of orbits actually takes place, and the desynchronization time of two points in Λ equals the symplectic area of the quadrangle two of whose corners are the points and whose edges are the (un)stable manifolds, see Lemma 8.2 of [9].

Thus the growth rate of the number prime periodic orbits γ with period T_γ satisfies an estimate reminiscent of the prime number theorem:

$$|\{\gamma \mid \exp(hT_\gamma) \le x\}| \sim \frac{x}{\ln x}$$

for a suitable $h > 0$, see Morita [13]. Of course – as Chapter 2 by V. Baladi on dynamical zeta functions shows – the similarity with number theory is no coincidence.

3. For the scattering problem the SRB (Sinai-Ruelle-Bowen) measure on Λ and not the Markov shift measure is the most important mixing measure (see Lopes and Markarian [11]).

4. It is interesting to see what happens if one drops the non-shadowing condition. For example, one may consider a one-parameter family of configurations of three disks, starting with all obstacles on a line and ending with a non-shadowing configuration.

The first configuration has exactly two bounded orbits, connecting the inner disk with the outer ones.

The parameter dependent birth of new bounded orbits is called the pruning phenomenon. The whole transition from regular to chaotic dynamics is described by Troll for a simplified model [21].

3.5 Irregular Scattering by Potentials

We planned to do scattering theory for potentials and ended up analysing bounded motion for obstacles. So it seems that we missed our aim. Of course this is not so. In this section we shall see how we can easily bring potentials into play. Then in Sect. 3.6 it will be explained how the set b of bounded orbits influences the set s of scattering orbits.

1) For simplicity we consider the three-disk-model, with the centres s_j of the disks. By a regularization argument we can take the characteristic function of the union $B = B_1 \cup B_2 \cup B_3$ of the disks as a potential to recover the above dynamics on the energy shell Σ. Instead, we consider potentials of the form

$$V(\mathbf{q}) := \hat{V}(\mathbf{q} - s_1) + \hat{V}(\mathbf{q} - s_2) + \hat{V}(\mathbf{q} - s_3) \qquad (3.22)$$

where $\hat{V}$ is a smooth centrally symmetric potential with profile $\tilde{V}(|\mathbf{q}|) = \hat{V}(\mathbf{q})$. We assume that $\tilde{V}$ is monotone decreasing and that $\tilde{V}(r) = 0$ for $r \geq r_0$. For simplicity we assume that the supports in (3.22) do not overlap ($R > 2r_0$) so that the particle is only under the influence of one of the three hills at a given time.

In order to analyse the Hamiltonian motion for (3.1), we may use the angular momenta

$$l_j(\mathbf{p}, \mathbf{q}) := \mathbf{p} \times (\mathbf{q} - s_j), \qquad (j = 1, 2, 3)$$

relative to the three hills. Then we can analyse the motion near one hill j using the auxiliary Hamiltonian function $H_j(\mathbf{p}, \mathbf{q}) := \frac{1}{2}|\mathbf{p}|^2 + \hat{V}(\mathbf{q} - s_j)$. H_j Poisson-commutes with l_j, and we thus can transform to polar coordinates w.r.t. the centre of the hill. If we assume that $\tilde{V}(0) > E$, (E being the energy of the particle), and that $\frac{d}{dr}\tilde{V}(r) < 0$ for $r \in (0, r_0)$ then each trajectory entering the support of the potential has exactly one pericentral point $\tilde{q}$ of minimal distance from s_j. We denote the angle $\angle(s_j - \tilde{q}, s_j)$ by ψ_j and use it together with l_i as a coordinate parametrizing the collision. Then the Poincaré maps

$$F_{j,k} : (l_j, \psi_j) \mapsto (l_k, \psi_k)$$

are area-preserving, and approximate the maps (3.12) for energy $E = \frac{1}{2}$ if the potential V approximates the characteristic function of B.

As one sees from this argument, many of the above statements carry over to motion in a potential.

2) Next we indicate how *Coulombic potentials* of the form (3.2) (possibly with an additional smooth potential W as discussed in Sect. 3.1) can be treated. For details we refer to the book [9].

The basic idea is to use techniques of Riemannian geometry after a regularization of the collision orbits.

In fact one can geometrize *any* motion in a smooth potential $V : M \to \mathbb{R}$ with Hamiltonian $H_{cl}(\mathbf{p}, \mathbf{q}) := \frac{1}{2}|\mathbf{p}|^2 + V(\mathbf{q})$. If the energy E of the particle is larger than $\sup_{\mathbf{q}} V(\mathbf{q})$, then the so-called *Jacobi metric*

$$g_{i,k}^E(\mathbf{q}) := (E - V(\mathbf{q})) \cdot \delta_{i,k}, \qquad (i, k = 1, \ldots, n, \mathbf{q} \in M)$$

is indeed a Riemannian metric on M. The geodesic motion in this metric is generated by the Hamiltonian function

$$H_E(\mathbf{p}, \mathbf{q}) := \frac{1}{2(E - V(\mathbf{q}))} |\mathbf{p}|^2, \qquad ((\mathbf{p}, \mathbf{q}) \in P).$$

So the energy shell $H_E^{-1}(1) \subset P$ coincides with the energy shell $H_{\text{cl}}^{-1}(E)$.

But for a regular value of the energy the *orbits* of a Hamiltonian system only depend on the form of the energy shell, not on the Hamiltonian function H of which it is a level set[6].

So the geodesics of the Jacobi metric g^E coincide with the energy E trajectories of H_{cl}, up to a time reparametrization.

However, for a Coulombic potential V of the form

$$V(\mathbf{q}) = \sum_{j=1}^{m} \frac{-Z_j}{|\mathbf{q} - \mathbf{s}_j|}$$

the Jacobi metric g^E on $M = \mathbb{R}_{\mathbf{q}}^2 - \{\mathbf{s}_1, \ldots, \mathbf{s}_m\}$ is geodesically incomplete: there are collision geodesics that meet the centres at $\mathbf{s}_j$ in finite time.

Nevertheless, we may regularize that motion by considering the branched covering surface

$$\mathbf{M} := \left\{ (q, Q) \in \mathbb{C} \times \mathbb{C} \,\middle|\, Q^2 = \prod_{l=1}^{n} (q - s_l) \right\} \tag{3.23}$$

of the configuration place $\mathbb{R}_{\mathbf{q}}^2 \cong \mathbb{C}$, with projection $(q, Q) \mapsto q$. It turns out that, after lifting g^E to $\mathbf{M}$ one can extend it smoothly to the branch points at $(s_j, 0) \in \mathbf{M}$. So on $\mathbf{M}$ geodesic motion becomes smooth and complete. Moreover, the Gaussian curvature turns out to be negative, and for $m \geq 3$ centres the surface $\mathbf{M}$ has handles.

Thus we are led to consider a geometric problem which is practically identical to the one considered by Hadamard in [6] – a fact which in the light of the introductory citation by Hadamard is somewhat ironic.

---

[6]The Hamiltonian vector field X_H of H satisfies the defining equation $\omega(X_H, \cdot) = dH$ with symplectic two-form ω. Since the tangent space of the energy shell at $x \in P$ coincides with the kernel of $dH(x)$, $dH(x)$ and thus also $X_H(x)$ are determined up to their lengths by that tangent space.

3.6 Time Delay and the Differential Cross Section

As a consequence of asymptotic completeness (see page 24) the probability that a celestial body is captured by our solar system is zero (neglecting collisions). However, it may stay a very long time before leaving, being repeatedly deflected by the sun and the planets. The notion of *time delay* quantifies this additional sojourn time.

In [14], Narnhofer gave a definition of that quantity for the case of a motion in a short-range potential V.

The time delay is the difference between the time spent by a particle in a configuration space ball of radius R, and the time spent by the corresponding incoming free particle, in the limit $R \to \infty$.

In order to include long-range potentials like the Coulombic potentials, in [9] that definition was slightly generalized:

Definition 3.8 *The* time delay $\tau : s \to \mathbb{R}$ *of a scattering state* $x \in s$ *is given by*

$$\tau(x) := \lim_{R \to \infty} \int_{\mathbb{R}} \big(\sigma(R) \circ \Phi^t(x) - \tag{3.24}$$

$$\tfrac{1}{2} \big(\sigma_\infty(R) \circ \Phi^t_\infty \circ \Omega^+_*(x) + \sigma_\infty(R) \circ \Phi^t_\infty \circ \Omega^-_*(x) \big) \big) \, dt,$$

where $\sigma(R) : P \to \{0, 1\}$ *and* $\sigma_\infty(R) : P_\infty \to \{0, 1\}$ *are the characteristic functions* $\sigma(R)(\mathbf{p}, \mathbf{q}) := \theta(R - |\mathbf{q}|)$ *and similarly for* $\sigma_\infty(R)$.

So the time delay is the difference between the time spent by the trajectory $\Phi^t(x)$ inside a ball $|\mathbf{q}| \leq R$ in the configuration space, and the mean of the times spent by the 'free' solutions $\Phi^t_\infty(\Omega^\pm_*(x))$ in the same ball, in the limit $R \to \infty$.

For short range potentials free motion is given by (3.3), whereas in the Coulombic case 'free' motion $\Phi^t_\infty : P_\infty \to P_\infty$ consists of the Kepler solutions generated by (3.4). So in that case physically speaking one compares scattering by the solar system with a situation where the sun has swallowed all planets.

It turns out ([9], Lemma 10.2) that time delay is a well-defined quantity which is constant on an orbit and invariant under time reversal.

One may also define time delay as a function τ^- of the initial data as described on page 24, that is, energy E, initial direction $\theta^- = \mathbf{p}^-/|\mathbf{p}^-| \in S^{n-1}$ and initial impact parameter $\mathbf{l}^-$. But then one has to take into account the measure zero subset of initial data which belong to semibounded orbits which are captured in the future. For these points τ^- equals ∞.

These exceptional trajectories belong to the stable manifold of the set b of bounded orbits. Thus they reflect the structure of this set. As an example, in the case of $m \geq 3$ Coulombic centres τ^- diverges on a Cantor set.

Next we consider the measure of the set of those scattering orbits of energy E whose time delay is larger than some positive time T. Therefore, we define

$$\kappa_E(T) := \int_{\mathbb{R}^{n-1}} \int_{S^{n-1}} \theta\big(\tau^-(E, \theta^-, \mathbf{l}^-) - T\big) \, d\theta^- \, d\mathbf{l}^-. \tag{3.25}$$

In the Coulombic case (which is typical) one finds the following behaviour (Thm. 10.6 of [9]):

- For $m = 1$ centre, the time delay is bounded $\tau(x) < C(H(x))$ by a function C of the energy which decreases to zero in the high energy limit (this statement is trivial if there is no additional smooth potential W).

- For $m \geq 2$ centres, there are constants $C_u > C_l > 0$ such that for $T \geq 1$ and E large the measure of the bounded orbits of energy E with time delay larger than T is bounded by

$$\exp\left(-C_u \sqrt{E} \ln(E) \cdot T\right) \leq \kappa_E(T) \leq \exp\left(-C_l \sqrt{E} \ln(E) \cdot T\right).$$

So in particular the probability to be captured for a time T decreases exponentially in T.

The last quantity we consider is the *differential cross section* defined on page 25.

In the presence of several obstacles or repelling hills of the potential like discussed in the last section, the differential cross section may show a Cantor set of 'rainbow singularities'.

However for attracting Coulomb potentials numerical calculations show nearly no difference between the Rutherford case (3.6) of $m = 1$ centre, the two-centre-case which is still integrable, and $m \geq 3$ centres, where irregular scattering occurs. In fact it can be shown that for arbitrary m, apart from the forward direction, the differential cross section is a smooth function in all these cases (Thm. 12.3 of [9]). Moreover it has the same forward asymptotics as the Rutherford cross section.

So in that case the classical dynamics looks rather regular as far as it is accessible in an experiment.

However, somewhat ironically the hidden irregular features of classical motion are expected to show up in the *quantum* cross section.

Without being able to go into details, the general idea is that divergences of the classical time delay connected with bound states of positive energy can give rise to quantum mechanical resonances.

In the simplest non-trivial case of the two-centre problem one has one bounded orbit of energy E which is reflected repeatedly at the two centres. This unstable closed orbit gives rise to families of scattering orbits of energy E with given initial and final directions θ^- and θ^+. Basically these orbits are indexed by their number of windings around the centres (see Thm. 12.1 of [9] for a more precise description).

Now asymptotically a scattering orbit with an additional winding has an additional length which converges to the length of the closed orbit. In a semiclassical picture each of the scattering orbits contributes to quantum scattering with a 'partial wave'. These waves *interfere*, and their phase difference is asymptotically proportional to the action $\oint \mathbf{p}\,\mathbf{dq}$ of the closed orbit. Depending on the energy E there may be constructive or destructive interference. In the two-centre case this gives rise to a regular pattern of quantum resonances.

In the case of a Cantor set of bounded orbits a result by Sjöstrand [18] indicates that the number of (physically relevant) resonances is governed by the fractional dimension of that set.

In the physics literature the quantum mechanics of scattering by three disks was considered by Gaspard and Rice [5] and by Cvitanović et al. [3], using periodic orbit theory.

Finally I would like to mention a connection with number theory. The scattering matrix for the Laplace-Beltrami operator in the modular domain is proportional to a quotient of Riemann zeta functions. It can be understood in terms of the geodesics which come from and go to the cusp [10]. The length differences of these geodesics can be described with the help of statistical mechanics.

Bibliography

[1] Berry, M.V.: Quantizing a Classically Ergodic System: Sinai's Billiard and the KKR Method. Annals of Physics **131**, 163–216 (1981)

[2] Eckhardt, B.: Irregular Scattering. Physica D **33**, 89–98 (1988)

[3] Cvitanović, P., Eckhardt, B., Russberg, G., Rosenqvist, P., Scherer, P.: Pinball Scattering. In: Quantum Chaos, G. Casati, B., Chirikov Eds. Cambridge University Press 1993

[4] Duistermaat, J.: Oscillatory Integrals, Lagrange Immersions and Unfolding of Singularities. Commun. Pure Appl. Math. **27**, 207–281 (1974)

[5] Gaspard, P., Rice, S.: Semiclassical quantization of the scattering from a classically chaotic repellor. Exact quantization of the scattering from a classically chaotic repellor. J. Chem. Phys. **90**, 2242–2254, 2255–2262 (1989) Erratum: J. Chem. Phys. **91**, 3280 (1989)

[6] Hadamard, J.: Les surfaces à courbures opposées et leur lignes géodesiques. Journ. Math. 5e série, t. 4, 27–73 (1898); Œvres, Tome II

[7] Hunziker, W.: Scattering in Classical Mechanics. In: Scattering Theory in Mathematical Physics. J.A. La Vita and J.-P. Marchand, Eds., Dordrecht: Reidel 1974

[8] Hunziker, W., Sigal, I.: The General Theory of N-Body Quantum Systems. CRM Proceedings and Lecture Notes **8**, 35–72, American Mathematical Society 1995

[9] Klein, M., Knauf, A.: Classical Planar Scattering by Coulombic Potentials. Lecture Notes in Physics m 13. Berlin, Heidelberg, New York: Springer; 1993

[10] Knauf, A.: Irregular Scattering, Number Theory, and Statistical Mechanics. In: Stochasticity and Quantum Chaos. Z. Haba et al, Eds. Dordrecht: Kluwer 1995

[11] Lopes, A., Markarian, R.: Open Billiards: Invariant and Conditionally Invariant Probabilities on Cantor Sets. SIAM J. Appl. Math. **56**, 651–680 (1996)

[12] Mather, J., McGehee, R.: Solutions of the Collinear four Body Problem which Become Unbounded in Finite Time. In: Dynamical Systems, Theory and Applications. J. Moser, Ed. Lecture Notes in Physics 38. Berlin, Heidelberg, New York: Springer; 1975

[13] Morita, T.: The symbolic representation of billiards without boundary condition. Trans. Am. Math. Soc. **325**, 819–828 (1991)

[14] Narnhofer, H.: Another Definition for Time Delay. Phys. Rev. D **22**, 2387–2390 (1980)

[15] Reed, M., Simon, B.: Methods of Modern Mathematical Physics. Vol. III: Scattering Theory. New York: Academic Press 1979

[16] Simon, B.: Wave Operators for Classical Particle Scattering. Commun. Math. Phys. **23**, 37–49 (1971)

[17] Sinai, Ya.: Dynamical Systems with Elastic Reflections. Russ. Math. Surv. **25**, 137–189 (1970)

[18] Sjöstrand, J.: Geometric Bounds on the Density of Resonances for Semiclassical Problems. Duke Math. J. **60**, 1–57 (1990)

[19] Smilansky, U.: The Classical and Quantum Theory of Chaotic Scattering. In: Chaos and Quantum Physics. M.-J. Giannoni et al, Eds. Les Houches LII. Amsterdam: North-Holland 1989

[20] Tél, T.: Transient Chaos. In: Directions in Chaos, Vol. 4 Ed. Hao Bai-lin. Singapore: World Scientific 1990

[21] Troll,G.: How to Escape a Sawtooth – Transition to Chaos in a Simple Scattering Model. Nonlinearity **5**, 1151–1192 (1992)

[22] Walters, P.: An Introduction to Ergodic Theory, New York, Heidelberg, Berlin: Springer 1982

Chapter 4

Quantum Chaos

Quantum chaos is defined to be the quantum mechanics for a classically chaotic – or, to be definite, ergodic – motion.

This definition first of all raises the question of how to compare classical and quantum mechanics. This problem is solved in the theory of pseudodifferential and Fourier integral operators, which grew out of the WKB method.

- In **classical mechanics** we have a *configuration space* $\mathbb{R}^n_q$ and a *phase space* $P := T^*\mathbb{R}^n_q \equiv \mathbb{R}^n_p \times \mathbb{R}^n_q$ of momenta $\mathbf{p}$ and positions $\mathbf{q}$ of the particle(s). Under suitable conditions Hamilton's equations

$$\dot{\mathbf{q}} = \frac{\partial H_{\mathrm{cl}}}{\partial \mathbf{p}}(\mathbf{p}, \mathbf{q}), \qquad \dot{\mathbf{p}} = -\frac{\partial H_{\mathrm{cl}}}{\partial \mathbf{q}}(\mathbf{p}, \mathbf{q})$$

of a smooth Hamiltonian function $H_{\mathrm{cl}} : P \to \mathbb{R}$ induces a flow $\Phi^t : P \to P$ on phase space which exists for all times $t \in \mathbb{R}$ and which maps the initial conditions $\mathbf{x}_0 = (\mathbf{p}_0, \mathbf{q}_0) \in P$ to the time-t-solution $\Phi^t(\mathbf{x}_0) = (\mathbf{p}(t), \mathbf{q}(t))$.

Example 1. $H_{\mathrm{cl}}(\mathbf{p}, \mathbf{q}) := \frac{1}{2}|\mathbf{p}|^2 + V(\mathbf{q})$. If the smooth potential V is bounded from below by a parabola, i.e,

$$V(\mathbf{q}) \geq -c\langle \mathbf{q}\rangle^2, \quad \text{with } \langle \mathbf{q}\rangle := \sqrt{1 + |\mathbf{q}|^2},$$

for some $c > 0$, then the velocity $\dot{\mathbf{q}} = \mathbf{p}$ is bounded by

$$\left|\frac{d}{dt}\langle \mathbf{q}\rangle\right| < |\dot{\mathbf{q}}| = \sqrt{2(E - V(\mathbf{q}))} \leq c'\langle \mathbf{q}\rangle$$

with energy $E := H(\mathbf{p}_0, \mathbf{q}_0)$ and $c' := \sqrt{2(|E| + c)}$. So $\langle \mathbf{q}(t)\rangle \leq e^{c'|t|}\langle \mathbf{q}_0\rangle$, the particle cannot escape to infinity in finite time and the flow Φ^t exists for all times.[1]

---

[1]Here we basically use $\langle \mathbf{q}\rangle$ instead of the function $|\mathbf{q}|$ to which it is asymptotic since $\langle \mathbf{q}\rangle \geq 1$ so that our inequalities have a simpler form. In the theory of pseudodifferential operators this function is useful since it is smooth.

For the *harmonic oscillator* the potential $V(\mathbf{q}) = \frac{1}{2}(\mathbf{q}, A\mathbf{q})$ with a positive $n \times n$-matrix A, and by a canonical transformation $(\mathbf{p}, \mathbf{q}) \mapsto (B\mathbf{p}, B\mathbf{q})$ into the orthonormal eigenbasis $(b_1, \ldots, b_n) = B$ of A we can assume w.l.o.g. that $V(\mathbf{q}) = \frac{1}{2}\sum_{j=1}^{n} \omega_j^2 q_j^2$ with frequencies $\omega_j > 0$. Then the flow Φ^t is given by

$$
\begin{aligned}
p_j(t) &= p_j(0)\cos(\omega_j t) - \omega_j q_j(0)\sin(\omega_j t), \\
q_j(t) &= \omega_j^{-1} p_j(0)\sin(\omega_j t) + q_j(0)\cos(\omega_j t).
\end{aligned}
$$

The closure of each orbit has thus the form of a $k \leq n$-dimensional torus. In the simplest case where all frequencies ω_j are rationally related, all orbits are closed.

- On the other hand, the arena of **quantum mechanics** is the Hilbert space $\mathcal{H} := L^2(\mathbb{R}_{\mathbf{q}}^n)$ of square integrable *wave functions* with inner product $(\varphi, \psi) := \int_{\mathbb{R}_{\mathbf{q}}^n} \overline{\varphi}(\mathbf{q})\psi(\mathbf{q})d\mathbf{q}$. The fact that the wave functions live on configuration space, not on phase space, is the basic reason for the difficulties one encounters if one wants to compare with the classical theory. We may go to momentum space $\mathbb{R}_{\mathbf{p}}^n$ by the unitary Fourier transformation

$$
\mathcal{F} : L^2(\mathbb{R}_{\mathbf{q}}^n) \to L^2(\mathbb{R}_{\mathbf{p}}^n),
$$

$$
(\mathcal{F}\varphi)(\mathbf{p}) := (2\pi\hbar)^{-n/2} \int_{\mathbb{R}_{\mathbf{q}}^n} \exp(-i\mathbf{p}\cdot\mathbf{q}/\hbar)\varphi(\mathbf{q})d\mathbf{q}
$$

with inverse $(\mathcal{F}^{-1}\psi)(\mathbf{p}) = (2\pi\hbar)^{-n/2} \int_{\mathbb{R}_{\mathbf{p}}^n} \exp(i\mathbf{q}\cdot\mathbf{p}/\hbar)\psi(\mathbf{p})d\mathbf{p}$, but clearly this does not solve the problem either.

The dynamics on $\mathcal{H}$ is given by the unitary one-parameter group

$$
U(t) := \exp(-itH_\hbar/\hbar) \in B(\mathcal{H}), \qquad (t \in \mathbb{R})
$$

generated by a self-adjoint operator $H_\hbar$ on $\mathcal{H}$. This maps the state $\psi_0 \in \mathcal{H}$ at time 0 to the time-t-state $\psi_t = U(t)\psi_0$.

Planck's constant $\hbar$ is in this context thought of as a positive parameter which in many cases is small compared to the other physical scales.

By Stone's theorem (see e.g. [20]) the unitarity of $U(t)$ is equivalent to the selfadjointness of $H_\hbar$, but in physical applications $H_\hbar$ is usually unbounded. Thus it cannot be defined on the whole Hilbert space $\mathcal{H}$ and we have to prove self-adjointness for a suitable dense domain of definition.

Example 2. A quantum particle in a potential V is described by the Hamiltonian

$$H_\hbar := -\tfrac{1}{2}\hbar^2 \Delta + V(\mathbf{q}).$$

If V is of the same class as the one considered in Example 1, then one can show (using similar arguments, see [15], Chapt. X.5) that $H_\hbar$ is essentially self-adjoint on $C_0^\infty(\mathbb{R}_q^n) \subset \mathcal{H}$. We now treat the harmonic oscillator and start with the case of $n = 1$ freedoms and frequency $\omega = 1$. Then

$$H_\hbar = \tfrac{1}{2}(\hat{p}^2 + \hat{q}^2) = A^* A + \frac{\hbar}{2} \quad \text{with } A := (\hat{q} + i\hat{p})/\sqrt{2},$$

$\hat{p} := -i\hbar \frac{d}{dq}$ and multiplication operator $\hat{q}$ by q. The ground state Ω_0 is annihilated by A and thus has the form $\Omega_0(q) = (\pi\hbar)^{-1/4} \exp(-q^2/2\hbar)$ (remember that you can find the normalization of the Gaussian on the 10 Mark bill!).

Moreover,

$$H_\hbar \Omega_m = \hbar(m + \tfrac{1}{2})\Omega_m \quad \text{with } \Omega_m := c_m \cdot (A^*)^m \Omega_0, \tag{4.1}$$

and $c_m = (\hbar^m m!)^{-1/2}$ since

$$[A, A^*] = \hbar \text{ and } [H, A^*] = \hbar A^*$$

for the commutator $[A, B] := AB - BA$.

We may write $\Omega_m(q) = (2^m m!)^{-\frac{1}{2}} P_m(q/\sqrt{\hbar})\Omega_0(q)$, since the m-th Hermite polynomial $P_m(q) = (-1)^m e^{q^2} \frac{d^m}{dq^m} e^{-q^2}$ satisfies the recurrence $\frac{d}{dq} P_m(q) = 2m P_{m-1}(q)$.

The spectrum $\text{spec}(H_\hbar)$ of the n-dimensional oscillator is obtained by addition:

$$\text{spec}(H_\hbar) = \left\{ \hbar(\sum_{j=1}^n \omega_j \cdot (m_j + \tfrac{1}{2})) \mid m_i \in \mathbb{N}_0 \right\}.$$

We see that the multiplicities of energy values are bounded (actually $= 1$) iff the frequencies are not rationally related, that is iff

$$\sum_{j=1}^n \omega_j \cdot m_j \neq 0 \qquad \text{for all } (m_1, \ldots, m_n) \in \mathbb{Z}^n \setminus \{0\}.$$

This is exactly the condition which guarantees that the generic classical invariant torus is n-dimensional.

The maximal degeneracy occurs if all frequencies are rationally related (or even equal) which corresponds to the classical situation where all orbits are closed.

This relation between (near)-degeneracies in the quantal spectrum and the existence of many closed orbits extends to a large class of Hamiltonians.

Now we want to compare

1. the states and

2. the time evolution

of classical and quantum mechanics. The heuristic correspondence principle of physics says that the two theories should make similar predictions in the *semi-classical limit* $\hbar \searrow 0$.

We first turn to the question of how to compare the states. The first answer which comes into mind is to compare the expectations of the classical and quantum *observables*. The classical observables are smooth phase space functions $O_{cl} : P \to \mathbb{R}$, whereas the quantum observables are self-adjoint operators $O_{\hbar}$ on $\mathcal{H}$. Whereas the classical expectation at time t equals $O_{cl} \circ \Phi^t(x_0)$, $x_0 \in P$, the quantum expectation equals

$$(\psi_t, O\psi_t) = (\psi_0, O_t \psi_0) \quad \text{with } O_t := U(-t)OU(t).$$

In particular we may compare the expectations of the classical phase space coordinates p_i, q_i, $i = 1, \ldots, n$ with the expectations of the multiplication operators $\hat{q}_j$ with q_j and the differential operators $\hat{p}_j := -i\hbar \frac{\partial}{\partial q_j}$ on $\mathcal{H}$.

Already on this level we have to assume that the wave function is in all the domains of definition of these operators (that is, decays fast enough at infinity and is smooth enough) in order to ensure that these quantum expectations are well-defined and finite.

We run into a new problem if we want to compare classical observables like dilation $\mathbf{q} \cdot \mathbf{p}$ with their quantum counterpart. By the commutation relation

$$[\hat{q}_j, \hat{p}_k] = i\hbar \delta_{j,k}$$

$\hat{\mathbf{p}} \cdot \hat{\mathbf{q}}$ is different from $\hat{\mathbf{q}} \cdot \hat{\mathbf{p}}$, and both operators are non-symmetric. So we should find some *quantization* prescription which attributes operators $O_{\hbar}$ to phase space functions O_{cl}.

4.1 Husimi Functions

> *"So our approximate eigenfunction u should be concentrated near x_0, while its Fourier transform û should be concentrated near ξ_0. The uncertainty principle tells us how well we can succeed in realizing the two conflicting goals."*
> C. Fefferman [7]

Leaving this question aside for a moment, we observe that we have not yet really solved the problem of comparing quantum with classical states, if we attribute to a wave function $\varphi \in \mathcal{H}$ the expectations of the positions $\hat{q}_j$ and momenta $\hat{p}_j$. The point is that we try to describe a function by $2n$ real numbers and thus loose too much information.

Instead we will associate to φ a phase space density ρ_φ, its Husimi function. Ideally we would like to interpret $\rho_\varphi(x)$ as the probability density of the quantum particle being localized at the phase space point $x \in P$. Now because of the uncertainty principle

$$\Delta \hat{p}_j \cdot \Delta \hat{q}_j \geq \tfrac{1}{2}\hbar \quad , \qquad (j = 1, \ldots, n) \tag{4.2}$$

(with standard deviation $\Delta O := \sqrt{(\varphi, O^2\varphi) - (\varphi, O\varphi)^2}$ of an observable O) a wave function $\varphi \in \mathcal{H}$ of norm $\|\varphi\| = 1$ cannot be localized in position *and* momentum so that this 'phase-space probability' is an ill-defined concept.

Mathematically this follows by the Cauchy-Schwarz inequality:
Since $\Delta O = \sqrt{(\varphi, \tilde{O}^2\varphi)}$ with $\tilde{O} := O - (\varphi, O\varphi)$,

$$\Delta \hat{p}_j \cdot \Delta \hat{q}_j \geq |(\tilde{p}_j\varphi, \tilde{q}_j\varphi)| \geq \mathrm{Im}(\tilde{q}_j\varphi, \tilde{p}_j\varphi) = \frac{1}{2i}(\varphi, [\tilde{q}_j, \tilde{p}_j]\varphi) = \tfrac{1}{2}\hbar. \tag{4.3}$$

On the other hand the position-momentum uncertainty becomes small in the semiclassical limit $\hbar \searrow 0$, so that in this limit the Husimi function ρ_ψ of a wave function ψ *can* be interpreted as a phase space probability. In order to define ρ_ψ we

1. start with a normed test function $\varphi \in \mathcal{H}$ concentrated at the origin of phase space P in the sense that its momentum and position expectations vanish $(\varphi, \hat{p}_j\varphi) = (\varphi, \hat{q}_j\varphi) = 0$.

2. Since we want to have test functions $\varphi_{\mathbf{p},\mathbf{q}}$ centered at arbitrary phase space points $(\mathbf{p}, \mathbf{q}) \in P$, we use the family $W_{\mathbf{p},\mathbf{q}} \in B(\mathcal{H})$ of unitary *Weyl operators*

$$\begin{aligned} W_{\mathbf{p},\mathbf{q}} &:= \exp\left(\frac{i}{\hbar}(\mathbf{p} \cdot \hat{\mathbf{q}} - \mathbf{q} \cdot \hat{\mathbf{p}})\right) \\ &= \exp\left(\frac{-i}{2\hbar}\mathbf{pq}\right) \exp\left(\frac{i}{\hbar}\mathbf{p} \cdot \hat{\mathbf{q}}\right) \exp\left(\frac{-i}{\hbar}\mathbf{q} \cdot \hat{\mathbf{p}}\right). \end{aligned}$$

These shift the position by $\mathbf{q}$, the momenta by $\mathbf{p}$, and commute up to a phase:

$$W_{\mathbf{p},\mathbf{q}}W_{\mathbf{p}',\mathbf{q}'} = \exp\left(\frac{i}{2\hbar}(\mathbf{p} \cdot \mathbf{q}' - \mathbf{q} \cdot \mathbf{p}')\right) W_{\mathbf{p}+\mathbf{p}',\mathbf{q}+\mathbf{q}'}.$$

We then introduce the family of wave functions $\varphi_{\mathbf{p},\mathbf{q}} := W_{\mathbf{p},\mathbf{q}}\varphi \in \mathcal{H}$,

$$\varphi_{\mathbf{p},\mathbf{q}}(\mathbf{Q}) = \exp\left(\frac{-i}{2\hbar}\mathbf{pq}\right) \exp(i\mathbf{p} \cdot \mathbf{Q}/\hbar)\varphi(\mathbf{Q} - \mathbf{q})$$

indexed by the phase space points $(\mathbf{p}, \mathbf{q}) \in P$. Their expectations are

$$(\varphi_{\mathbf{p},\mathbf{q}}, \hat{p}_j\varphi_{\mathbf{p},\mathbf{q}}) = p_j \quad , \quad (\varphi_{\mathbf{p},\mathbf{q}}, \hat{q}_j\varphi_{\mathbf{p},\mathbf{q}}) = q_j.$$

3. Then we associate with an arbitrary wave function $\psi \in \mathcal{H}$ which we want to analyze its *Husimi function* $\rho_\psi : P \to \mathbb{R}$,

$$\rho_\psi(\mathbf{p}, \mathbf{q}) := (2\pi\hbar)^{-n} |(\varphi_{\mathbf{p},\mathbf{q}}, \psi)|^2 . \qquad (4.4)$$

We see that $\rho_\psi \geq 0$ and

$$\int_P \rho_\psi(\mathbf{p}, \mathbf{q}) \, d\mathbf{p} d\mathbf{q} = 1.$$

So ρ_ψ is a probability density on phase space. Although a *probabilistic interpretation* of $\rho_\psi(\mathbf{p}, \mathbf{q})$ as the probability of finding the quantum particle at the phase space point $(\mathbf{p}, \mathbf{q})$ fails, it becomes true in the semiclassical limit, if our test functions $\varphi_{\mathbf{p},\mathbf{q}}$ have the minimal possible uncertainty, that is, if we choose them in a way that the uncertainty inequalities (4.2) become equalities.

Wave functions with that property are called *coherent states*. In fact it suffices to consider φ, since the $\varphi_{\mathbf{p},\mathbf{q}}$ are coherent if φ is.

By the first inequality in (4.3) for a coherent state φ the functions $\hat{p}_j\varphi$ and $\hat{q}_j\varphi$ must be linearly dependent, and by the second inequality the proportionality factor must be imaginary: $\hat{p}_j\varphi = ia_j\hat{q}_j\varphi$ with $a_j > 0$. This means that $\varphi \in \mathcal{H}$ is a Gaussian:

$$\varphi(\mathbf{Q}) = c \cdot \exp\left(-\sum_{j=1}^{n} a_j Q_j^2 / 2\hbar\right).$$

If we want all the uncertainties to be equal ($\Delta\hat{p}_j = \Delta\hat{q}_k$ for $1 \leq j, k \leq n$), then we must set $a_j := 1$. Then

$$\varphi_{\mathbf{p},\mathbf{q}}(\mathbf{Q}) = \exp\left(\frac{-i}{2\hbar}\mathbf{p}\mathbf{q}\right)(\pi\hbar)^{-n/4}\exp\left(\left(i\mathbf{p}\mathbf{Q} - (\mathbf{Q}-\mathbf{q})^2/2\right)/\hbar\right), \qquad (4.5)$$

see Figure 4.1. One should think of $\varphi_{\mathbf{p},\mathbf{q}}$ as a quantum analog of the phase space point $(\mathbf{p}, \mathbf{q})$.

In particular by (4.1) $\varphi_{0,0} \equiv \varphi$ is the ground state of the n-dim. harmonic oscillator with frequencies $\omega_i = 1$: $\varphi(\mathbf{Q}) = \Omega_0(Q_1) \cdot \ldots \cdot \Omega_0(Q_n)$.

Example 3. Using the last remark, we calculate the Husimi functions ρ_m for the $m+1$-th eigenfunction Ω_m of the one-dim. harmonic oscillator, see Figure 4.2.

By (4.1) $2\pi\hbar\rho_m(\mathbf{p}, \mathbf{q})$ equals the absolute square of

$$(\Omega_m, W_{\mathbf{p},\mathbf{q}}\Omega_0)$$

$$= c_m\left((A^*)^m\Omega_0, \exp\left(\frac{i}{\hbar}(\mathbf{p}\cdot\hat{\mathbf{q}} - \mathbf{q}\cdot\hat{\mathbf{p}})\right)\Omega_0\right)$$

$$= c_m\left((A^*)^m\Omega_0, \exp\left(\frac{i}{\sqrt{2\hbar}}[(\mathbf{p} - i\mathbf{q})\cdot A^* + (\mathbf{p} + i\mathbf{q})\cdot A]\right)\Omega_0\right)$$

$$= c_m\exp(-(\mathbf{p}^2 + \mathbf{q}^2)/4\hbar) \cdot$$

$$\left(\exp\left(\frac{-i}{\sqrt{2\hbar}}(\mathbf{p} + i\mathbf{q})\cdot A\right)(A^*)^m\Omega_0, \exp\left(\frac{i}{\sqrt{2\hbar}}(\mathbf{p} + i\mathbf{q})\cdot A\right)\Omega_0\right)$$

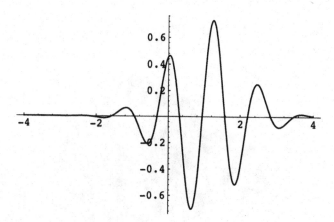

Figure 4.1: Real part of coherent state $\varphi_{\mathbf{p},\mathbf{q}}$ centered at $\mathbf{p} = 5$, $\mathbf{q} = 1$, with $\hbar = 1$.

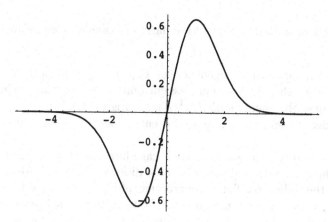

Figure 4.2: Second eigenfunction Ω_1 of the harmonic oscillator.

$$
\begin{aligned}
&= c_m \exp(-(\mathbf{p}^2 + \mathbf{q}^2)/4\hbar) \cdot \\
&\quad \left(\left(\mathbf{A}^* + \frac{\mathbf{q} - i\mathbf{p}}{\sqrt{2}}\right)^m \exp\left(\frac{-i(\mathbf{p} + i\mathbf{q})}{\sqrt{2\hbar}} \cdot \mathbf{A}\right) \Omega_0, \exp\left(\frac{i(\mathbf{p} + i\mathbf{q})}{\sqrt{2\hbar}} \cdot \mathbf{A}\right) \Omega_0 \right) \\
&= (m!)^{-1/2} \exp(-(\mathbf{p}^2 + \mathbf{q}^2)/4\hbar) \left(\frac{\mathbf{q} + i\mathbf{p}}{\sqrt{2\hbar}}\right)^m,
\end{aligned}
$$

since $\mathbf{A}\Omega_0 = 0$.

Thus the Husimi function has the form

$$
\rho_m(\mathbf{p}, \mathbf{q}) = (2\pi\hbar)^{-1}((\mathbf{p}^2 + \mathbf{q}^2)/2\hbar)^m \exp(-(\mathbf{p}^2 + \mathbf{q}^2)/2\hbar)/m! \qquad (4.6)
$$

and depends only on the distance from the origin in the phase space plane P, see Figure 4.3.

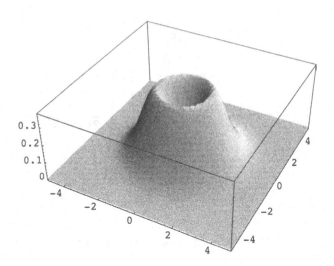

Figure 4.3: Husimi function ρ_1 of the harmonic oscillator eigenfunction Ω_1.

Moreover if we consider eigenenergies $E_m = (\frac{1}{2} + m)\hbar$ near $E \geq 0$ by set-ting $\hbar_m := E/m$, then in the semiclassical limit $m \to \infty$ the Husimi function concentrates near the energy shell $H_{cl}^{-1}(E)$, see Figure 4.4.

In this one-dimensional example this energy shell consists of just one closed orbit.

Now one may by the same method calculate the Husimi function for frequencies of the harmonic oscillator which are different from one. Also, the Husimi functions for the n-dim. oscillator are products of the above one-dim. expressions.

So these phase space densities concentrate in the semiclassical limit to the invariant probability measures on the minimal $k \leq n$-dim. invariant tori of the classical flow.

In order to catch this kind of concentration phenomena, the notion of the *frequency set* is useful. This is the subset $FS(\psi_\hbar)$ of the phase space P on which a family $\hbar \mapsto \psi_\hbar \in \mathcal{H}$ of normed wave functions concentrates in the semiclassical limit.

We define it through its complement $P - FS(\psi_\hbar)$.

Definition 4.1 *A point* $(\mathbf{p}, \mathbf{q}) \in P$ *does* **not** *belong to the* frequency set of $\psi_\hbar$ *$((\mathbf{p}, \mathbf{q}) \in P - FS(\psi_\hbar))$ iff there exists a neighbourhood $U \subset P$ of it on which the Husimi functions $\rho_{\psi_\hbar}$ are uniformly $\mathcal{O}(\hbar^\infty)$, that is, for all $k > 0$ there exists a c_k with*

$$\rho_{\psi_\hbar}|_U \leq c_k \cdot \hbar^k.$$

Example 4. 1) The coherent states $\varphi_{\mathbf{p}, \mathbf{q}}$ are normed and depend on the parameter $\hbar$. By formula (4.6) for $\rho_0 = \rho_{\varphi_{0,0}}$ their Husimi Function is a Gaussian in P centered

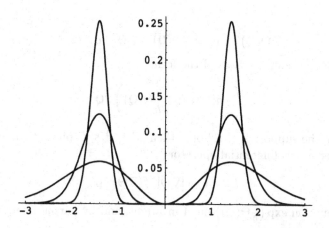

Figure 4.4: Husimi functions $\rho_m(0, \mathbf{q})$, with $m = 1$, $m = 4$ and $m = 16$, and $\hbar = 1/m$.

at $(\mathbf{p}, \mathbf{q})$:

$$\rho_{\varphi_{\mathbf{p},\mathbf{q}}}(\mathbf{P}, \mathbf{Q}) = (2\pi\hbar)^{-n} \exp(-((\mathbf{P} - \mathbf{p})^2 + (\mathbf{Q} - \mathbf{q})^2)/2\hbar).$$

So their frequency set consists of that point:

$$\mathrm{FS}(\varphi_{\mathbf{p},\mathbf{q}}) = \{(\mathbf{p}, \mathbf{q})\}.$$

2) Whereas the above frequency set could be determined by an explicit Gaussian integration, the statement also follows from a non-stationary phase argument which we will apply now.

The frequency set $\mathrm{FS}(\psi_\hbar)$ of the family $\hbar \mapsto \psi_\hbar \in \mathcal{H}$,

$$\psi_\hbar(\mathbf{Q}) := a(\mathbf{Q}) \exp\left(\frac{i}{\hbar}\theta(\mathbf{Q})\right)$$

with a function $a \in C_0^\infty(\mathbb{R}_\mathbf{q}^n)$ of compact support is contained in the Lagrange manifold[2]

$$\{(\mathbf{Q}, \nabla\theta(\mathbf{Q})) \in P \mid \mathbf{Q} \in \mathrm{supp}(a)\}. \tag{4.7}$$

Namely by definition $(2\pi\hbar)^n \cdot \rho_{\psi_\hbar}(\mathbf{p}, \mathbf{q})$ is the absolute square of

$$(\psi_\hbar, \varphi_{\mathbf{p},\mathbf{q}})$$

$$= (\pi\hbar)^{-n/4} e^{-i\mathbf{p}\mathbf{q}/2\hbar} \int_{\mathbb{R}_\mathbf{q}^n} a(\mathbf{Q}) \exp\left(\frac{-i}{\hbar}\theta(\mathbf{Q})\right) \exp\left(\frac{i}{\hbar}\mathbf{p} \cdot \mathbf{Q} - (\mathbf{Q} - \mathbf{q})^2/2\hbar\right) d\mathbf{Q}$$

$$= (\pi\hbar)^{-n/4} e^{-i\mathbf{p}\mathbf{q}/2\hbar} \int_{\mathbb{R}_\mathbf{q}^n} a(\mathbf{Q}) \exp\left(\frac{i}{\hbar}\Theta(\mathbf{Q})\right) d\mathbf{Q}$$

---

[2]Here the property $\partial p_j/\partial q = \partial p_k/\partial q_j$ of a Lagrange manifold reduces to the commutativity of partial derivatives of θ.

with

$$\Theta(\mathbf{Q}) := \mathbf{p} \cdot \mathbf{Q} - \theta(\mathbf{Q}) + i(\mathbf{Q} - \mathbf{q})^2/2. \tag{4.8}$$

So we have to consider integrals of the form

$$\int_{\mathbb{R}^n} a(\mathbf{Q}) \exp\left(\frac{i}{\hbar}\Theta(\mathbf{Q})\right) d\mathbf{Q}.$$

If $\nabla\Theta \neq 0$ on the support $A \subset \mathbb{R}^n$ of a, then $|\nabla\Theta|^{-1} \leq C$ on that compact region. Then the first order differential operator

$$L := \chi_A \cdot |\nabla\Theta|^{-2}(\nabla\Theta) \cdot \hat{\mathbf{p}}$$

leaves the function $\exp\left(\frac{i}{\hbar}\Theta(\mathbf{Q})\right)$ on A invariant, and by partial integration

$$\left|\int_{\mathbb{R}^n} a(\mathbf{Q}) \exp\left(\frac{i}{\hbar}\Theta(\mathbf{Q})\right) d\mathbf{Q}\right| = \hbar \left|\int_{\mathbb{R}^n} b(\mathbf{Q}) \exp\left(\frac{i}{\hbar}\Theta(\mathbf{Q})\right) d\mathbf{Q}\right|$$

with $b := |\nabla\Theta|^{-2}\nabla\Theta \cdot \nabla a$. By iteration we see that the integral is $\mathcal{O}(\hbar^\infty)$.

For Θ of the above form (4.8) $\nabla\Theta$ vanishes only if $\mathbf{q} = \mathbf{Q}$ and $\mathbf{p} = \nabla\Theta(\mathbf{q})$. So we proved (4.7).

Families $\psi_\hbar$ of the above form appear in the WKB approximation of wave functions.

4.2 Pseudodifferential Operators

> "... it seems to me that there has been in the literature
> entirely too much emphasis on quantization, (i.e. general
> methods of obtaining quantum mechanics from classical methods)
> as opposed to the converse problem of the classical limit of
> quantum mechanics. This is unfortunate since the latter is an
> important question for various areas of modern physics
> while the former is, in my opinion, a chimera."
> B. Simon [19]

The concepts of Husimi functions and the frequency set answer our first question on how to compare classical and quantal states.

However, we are particularly interested in *eigen*states of a Hamiltonian operator and their classical counterparts, the probability measures on phase space which are left invariant by the flow of the Hamiltonian function. Thus we need a prescription of how to *quantize* a phase space function.

If that function is sufficiently well-behaved, the resulting operator is a so-called *pseudodifferential operator*, and the unitary time evolution generated by that operator is a *Fourier integral operator*.

In this section we give a short introduction into the theory of pseudodifferential operators, whereas Fourier integral operators are discussed in Section 4.3.

Then the techniques will be applied to quantum chaos. Of course, this introduction can only be extremely superficial, and we shall as far as possible suppress technical problems.

We recommend the book [16] by Robert and the Volumes [11] by Hörmander to the interested reader. Helffer recently gave a beautiful untechnical introduction into the subject [9], containing an extensive bibliography.

A *quantization* is a linear map $F \mapsto F^Q$ from phase space functions $F : P \to \mathbb{R}$ to operators F^Q on $\mathcal{H}$ which should ideally extend our quantization prescriptions $p_j \mapsto \hat{p}_j$, $q_j \mapsto \hat{q}_j$ for the momenta and positions and convert the Poisson bracket

$$\{F, G\} := \sum_{j=1}^n \left(\frac{\partial F}{\partial p_j} \frac{\partial G}{\partial q_j} - \frac{\partial F}{\partial q_j} \frac{\partial G}{\partial p_j} \right)$$

into the commutator:

$$[F^Q, G^Q] = \frac{\hbar}{i} \{F, G\}^Q. \tag{4.9}$$

In fact it can be proved (see [1], Theorem 5.4.9) that these aims are, strictly speaking, incompatible. However the violation of (4.9) will become small in the semiclassical limit $\hbar \searrow 0$.

One quantization prescription is based on the coherent states which we considered in the last section. This so-called *Anti-Wick* quantization is given by

$$F^{\mathrm{AW}} := (2\pi\hbar)^{-n} \int_P F(\mathbf{p}, \mathbf{q}) \Pi_{\mathbf{p}, \mathbf{q}} d\mathbf{p} d\mathbf{q},$$

where $\Pi_{\mathbf{p}, \mathbf{q}}$ is the orthogonal projection on the coherent state $\varphi_{\mathbf{p}, \mathbf{q}}$. Of course one has to consider domain questions for F^{AW} if F is unbounded.

A simple calculation shows that $p_j^{\mathrm{AW}} = \hat{p}_j$ and $q_j^{\mathrm{AW}} = \hat{q}_j$. However, general configuration space functions V are not converted into the multiplication operators with these functions, but with a function $\tilde{V}$:

$$\begin{aligned}
\tilde{V}(\mathbf{Q}) &= (\pi\hbar)^{-n/2} \int_{\mathbb{R}_{\mathbf{q}}^n} e^{-(\mathbf{Q}-\mathbf{q})^2/\hbar} V(\mathbf{q}) d\mathbf{q} \\
&= (2\pi)^{-n/2} \int_{\mathbb{R}_{\mathbf{q}}^n} e^{-\mathbf{u}^2/2} V\left(\mathbf{Q} - \sqrt{\hbar/2}\mathbf{u}\right) d\mathbf{u} \\
&= V(\mathbf{Q}) + \frac{\hbar}{4} \Delta V(\mathbf{Q}) + \mathcal{O}(\hbar^2).
\end{aligned}$$

Although the difference between $\tilde{V}$ and V becomes small in the semiclassical limit, it is sometimes unwanted.

On the other hand the Anti-Wick quantization has the nice property that for positive phase space functions F the expectation are positive, too:

$$(\psi, F^{\mathrm{AW}}\psi) = (2\pi\hbar)^{-n} \int_P F(\mathbf{p}, \mathbf{q})(\psi, \Pi_{\mathbf{p}, \mathbf{q}}\psi) d\mathbf{p} d\mathbf{q} \geq 0.$$

There are several other quantization prescriptions, each with its merits and draw-backs, and the general attitude is to change freely between them in order to obtain the best results.

One such quantization which we shall use is the *Weyl quantization* $F \mapsto F^W$ with the action

$$(F^W \psi)(\mathbf{Q}) := (2\pi\hbar)^{-n} \int_P \exp\left(\frac{i}{\hbar}(\mathbf{Q}-\mathbf{q}) \cdot \mathbf{p}\right) F(\mathbf{p}, \tfrac{1}{2}(\mathbf{Q}+\mathbf{q}))\psi(\mathbf{q})d\mathbf{p}d\mathbf{q} \quad (4.10)$$

of the operator F^W on a wave function ψ.

By the *Plancherel formula*

$$\psi(\mathbf{Q}) = (2\pi\hbar)^{-n} \int_P \exp\left(\frac{i}{\hbar}(\mathbf{Q}-\mathbf{q}) \cdot \mathbf{p}\right) \psi(\mathbf{q})d\mathbf{p}d\mathbf{q}$$

of Fourier theory coordinate functions V are converted into multiplication oper-ators V^W with V, and using the same formula one sees that polynomials in the momenta p_j go into the same polynomials in the differential operators $\hat{p}_j$. In par-ticular the Schrödinger operator

$$-\tfrac{1}{2}\hbar^2\Delta + V(\mathbf{q}) = (\tfrac{1}{2}\mathbf{p}^2 + V(\mathbf{q}))^W. \quad (4.11)$$

Moreover, the strange-looking convex combination $\tfrac{1}{2}(\mathbf{Q}+\mathbf{q})$ appearing in the def-inition (4.10) makes the operator F^W symmetric for real F, even if that phase space function contains mixed $\mathbf{p}, \mathbf{q}$ terms.

We shall consider the above quantizations of phase space functions which are in the *symbol class* $S^m(P)$, $m \in \mathbb{Z}$, of smooth phase space functions F whose derivatives satisfy the following inequalities[3]

$$\forall(\alpha,\beta) \in \mathbb{N}_0^n \times \mathbb{N}_0^n \ \exists C_{\alpha,\beta} > 0: \ \left|\partial_{\mathbf{p}}^\alpha \partial_{\mathbf{q}}^\beta F(\mathbf{p},\mathbf{q})\right| \leq C_{\alpha,\beta} \langle(\mathbf{p},\mathbf{q})\rangle^{m-|\alpha|-|\beta|}.$$

Obviously they are related by $S^m(P) \subset S^{m+1}(P)$.

Remark 4.2 *The phase space coordinates p_j, q_j and the Hamiltonian function of the harmonic oscillator and its local perturbations fall into the class $S^2(P)$. One reason for the importance of the symbol class $S^2(P)$ is that potentials which fall off faster than $-\mathbf{q}^2$ may lead the particle to infinity in finite time.*

For simplicity of the argument we will mostly consider symbols F in $S^0(P)$. These phase space functions are bounded together with all its derivatives. The Calderon-Vaillancourt Theorem (Thm. II–38 of [16]) then implies that their Weyl quantizations F^W are bounded operators on $\mathcal{H}$. Although most Hamiltonians and observables of quantum mechanics are unbounded, it is often possible to localize a problem using a partition of unity in P and effectively working with bounded operators.

---

[3]Here we use multi-index notation $\partial_{\mathbf{p}}^\alpha := \partial_{p_1}^{\alpha_1} \dots \partial_{p_n}^{\alpha_n}$, $|\alpha| := \sum_{j=1}^n \alpha_j$, $\alpha! := \prod_{j=1}^n \alpha_j$ etc.

Moreover, the above function classes $S^m(P)$ are by no means exhaustive. As an example, they do not even contain a bounded periodic potential like $V(q) = \cos(q)$. In such a case one has quite some freedom in the semiclassical theory to define other symbol classes which are taylor-made for the analysis of a given operator.

Exercise. *Prove the formula*

$$F^{\mathrm{AW}} = \tilde{F}^{\mathrm{W}} \tag{4.12}$$

with the smoothed symbol

$$\tilde{F}(\mathbf{P}, \mathbf{Q}) := (\pi\hbar)^{-n} \int_P \exp\left(-\left[(\mathbf{P} - \mathbf{p})^2 + (\mathbf{Q} - \mathbf{q})^2\right]/\hbar\right) F(\mathbf{p}, \mathbf{q}) d\mathbf{p} d\mathbf{q}$$

which connects the Weyl and the Anti-Wick quantizations.
Show that $\tilde{F} - F = \mathcal{O}(\hbar)$ for $F \in S^0(P)$.

The next step is to multiply operators of the above form. Obviously the product operator cannot in general have a symbol which is just the product of the symbols of the factors (otherwise they would have to commute).

Instead, that symbol $\sigma\left(F^{\mathrm{W}} G^{\mathrm{W}}\right)$ will depend on $\hbar$. So if we want to stay in our symbol space, we have to enlarge it. Thus we introduce $\hbar$-*admissible symbols* in $S^m_\hbar(P)$ which are functions $F : P \times (0, 1] \to \mathbb{C}$ of the phase space point and $\hbar$ with expansions in $\hbar$ of the form

$$F \sim \sum_{j=0}^{\infty} F_j \hbar^j \quad \text{with } F_j \in S^{m-j}(P).$$

The $\sim$ symbol means that for large enough N

$$F - \sum_{j=0}^{N} F_j \hbar^j = \mathcal{O}(\hbar^{N+1}) \text{ in } S^0(P).$$

The quantization F^{W} of such a function is then a $\hbar$-dependent operator. An operator which can be written in that form is called $\hbar$-*pseudodifferential operator*.

The symbol F of such an operator is also denoted by $\sigma(F^{\mathrm{W}})$. Its leading function F_0 is the *principal symbol*, and F_1 is called *subprincipal*.

Example. For Schrödinger operators $H_\hbar$ of the form (4.11) $\sigma(H_\hbar)(\mathbf{p}, \mathbf{q}) = \frac{1}{2}\mathbf{p}^2 + V(\mathbf{q})$ so that the symbol is principal.

However, if the potential V is $\mathcal{L}$-periodic with some lattice $\mathcal{L}$, then it is useful to consider the family of operators

$$H_\hbar(\mathbf{k}) := \frac{1}{2}\hbar(-i\nabla + \mathbf{k})^2 + V \quad \text{on } L^2(\mathbb{R}^n/\mathcal{L}),$$

indexed by the wave vector $\mathbf{k}$ which takes values in the Brillouin zone $(\mathbb{R}^n)^*/\mathcal{L}^*$, $\mathcal{L}^*$ denoting the dual lattice. Then the principal symbol equals $\frac{1}{2}\mathbf{p}^2 + V(\mathbf{q})$, the subprincipal symbol has the form $\mathbf{p} \cdot \mathbf{k}$, and $\sigma(H_\hbar)(\mathbf{p}, \mathbf{q})(\mathbf{k}) = \frac{1}{2}(\mathbf{p} + \hbar\mathbf{k})^2 + V(\mathbf{q})$.

By a linear change of variables one proves that the product $F^W G^W$ of two Weyl-quantized operators with $F, G \in S_\hbar^0(P)$ is a $\hbar$-pseudodifferential operator with symbol

$$\sigma\left(F^W G^W\right)(\mathbf{p}, \mathbf{q})$$
$$= \exp\left(\tfrac{1}{2} i\hbar (D_{\mathbf{q}^I} D_{\mathbf{p}^{II}} - D_{\mathbf{q}^{II}} D_{\mathbf{p}^I})\right) F(\mathbf{p}^I, \mathbf{q}^I) G(\mathbf{p}^{II}, \mathbf{q}^{II})\big|_{\substack{\mathbf{p}^I = \mathbf{p}^{II} = \mathbf{p} \\ \mathbf{q}^I = \mathbf{q}^{II} = \mathbf{q}}}.$$

By evaluating the first two terms in the series of the exponential function, we see that the principal symbol equals the product $F_0 G_0$, and that the subprincipal symbol has the form

$$F_0 G_1 + F_1 G_0 - \frac{i}{2}\{F_0, G_0\}.$$

Thus up to higher orders in $\hbar$ Weyl quantization satisfies the desideratum (4.9) for the relation between commutator and Poisson bracket.

Formula (4.12), which connects Weyl- and Anti-Wick quantization, implies that for $F \in S_\hbar^0(P)$ the operator F^{AW}, too is a $\hbar$-pseudodifferential operator. Moreover, by Gaussian integration one sees that $\hbar^{-1}(F - \tilde{F}) \in S_\hbar^0(P)$ so that by the Calderon-Vaillancourt Theorem the difference

$$\|F^{AW} - F^W\|_{B(\mathcal{H})} = \mathcal{O}(\hbar) \tag{4.13}$$

between the quantizations vanishes in the semiclassical limit.

Differential operators F^W cannot increase the *support* of wave functions:

$$\mathrm{supp}(F^W \psi) \subset \mathrm{supp}(\psi), \qquad (\psi \in \mathcal{H}). \tag{4.14}$$

As the name suggests, not all pseudodifferential operators are differential, a typical example being the resolvent

$$\left(F^W - E\hat{\mathbb{1}}\right)^{-1}, \qquad E \notin \mathrm{spec}(F^W)$$

of a Schrödinger operator. *Pseudo*differential operators F^W can increase the support (a simple example is the resolvent $1/(-\hbar^2\Delta + \hat{\mathbb{1}})$ of the Laplacian in $\mathbb{R}^3$ with the strictly positive kernel $\frac{e^{-|x-y|/\hbar}}{4\pi\hbar^2|x-y|}$.)

Nevertheless they share the property not to increase the *frequency set* of a family $\hbar \mapsto \psi_\hbar \in \mathcal{H}$,:

$$\mathrm{FS}\left(F^W \psi_\hbar\right) \subset \mathrm{FS}\left(\psi_\hbar\right).$$

This can be seen using Anti-Wick quantization: Up to the constant $(2\pi\hbar)^{-n}$ the Husimi function of $F^{AW}\psi_\hbar$ is the absolute square of

$$\left(\psi_\hbar, F^{AW}\varphi_{\mathbf{p}, \mathbf{q}}\right)$$
$$= (2\pi\hbar)^{-n}\left(\psi_\hbar, \int_P F(\mathbf{p}', \mathbf{q}')\Pi_{\mathbf{p}', \mathbf{q}'}\varphi_{\mathbf{p}, \mathbf{q}}\, d\mathbf{p}'d\mathbf{q}'\right)$$

$$= (2\pi\hbar)^{-n} \int_P F(\mathbf{p}',\mathbf{q}') \, (\psi_\hbar, \varphi_{\mathbf{p}',\mathbf{q}'}) \, (\varphi_{\mathbf{p}',\mathbf{q}'}, \varphi_{\mathbf{p},\mathbf{q}}) \, d\mathbf{p}' d\mathbf{q}'$$

$$= (2\pi\hbar)^{-n} \int_P F(\mathbf{p}',\mathbf{q}') \cdot (\psi_\hbar, \varphi_{\mathbf{p}',\mathbf{q}'}) \cdot$$
$$\cdot \exp\left(-\left[(\mathbf{q}-\mathbf{q}')^2 + (\mathbf{p}-\mathbf{p}')^2 - 2i(\mathbf{q}'\mathbf{p}-\mathbf{p}'\mathbf{q})\right]/4\hbar\right) d\mathbf{p}' d\mathbf{q}'.$$

Thus if $(\mathbf{p},\mathbf{q}) \notin \mathrm{FS}(\psi_\hbar)$, then the second or the third factor in the integral is $\mathcal{O}(\hbar^\infty)$.

One defines functions $\chi(O)$ of self-adjoint operators O using the *functional calculus* (see [15], Vol. I). This is also possible in the semi-classical context.

If $H \in S_\hbar^2(P)$ is a real, positive symbol, then H^W can be proven to be an essentially self-adjoint operator [16], Thm. III–4. As an example we can treat a local perturbation H of the harmonic oscillator.

Now for $\chi \in C_0^\infty(\mathbb{R})$ the operator $\chi(H^W)$ defined by functional calculus is a $\hbar$-pseudodifferential operator with principal symbol $\chi(H)$. More precisely for $H_{cl} \in S^2(P)$ the phase space functions g_j appearing in the symbol

$$\sigma\left(\chi(H_{cl})^W)\right) \sim \sum_{j=0}^\infty g_j \hbar^j$$

are supported in $\mathrm{supp}(\chi(H_{cl})) \subset P$.

This is a starting observation of the proof that *eigenfunctions are semi-classically localized near their energy shell*.

First of all, eigenfunctions $\psi_\hbar$ satisfying the eigenequation

$$H^W \psi_\hbar = E_\hbar \psi_\hbar$$

are square integrable ($\psi_\hbar \in \mathcal{H}$) and thus in some sense localized. Thus in order to exclude the presence of essential spectrum, the corresponding classical object, the *energy shell*

$$H_{cl}^{-1}(E) \subset P$$

of the principal symbol H_{cl} should also meet some localization assumption.

More precisely we assume that the thickened energy shell $H_{cl}^{-1}([E - 2\varepsilon, E + 2\varepsilon])$ is compact. Then the spectrum

$$\mathrm{spec}\left(H^W\right) \cap [E - \varepsilon, E + \varepsilon]$$

in the ε-neighbourhood of E is *purely discrete*, i.e., consists of isolated eigenvalues with finite multiplicity[4].

Actually it suffices to assume that the thickened energy shell has finite volume, see [16], Thm. III-13.

---

[4]In that kind of theorems one always has to assume that $\hbar$ is small enough, since already the subprincipal symbol $\hat{\mathbb{1}}$ shifts the spectrum by $\hbar$.

Now for simplicity of the argument we consider only eigenfunctions $\psi_{\hbar_m}$ with m-independent eigenvalue E (for a Schrödinger operator H^W and $E > V_{\min}$ we can always find a decreasing sequence $m \mapsto \hbar_m \searrow 0$ and eigenfunctions $\psi_{\hbar_m}$ with common eigenvalue E, using the Rayleigh-Ritz principle, Thm. XIII.3 of [15]). Then for that family $m \mapsto \psi_{\hbar_m}$ of normed eigenfunctions the frequency set is localized in the energy shell:

$$\mathrm{FS}(\psi_{\hbar_m}) \subset H_{\mathrm{cl}}^{-1}(E). \tag{4.15}$$

To see this, we take a cutoff function $\chi \in C_0^\infty(\mathbb{R})$ which equals one on the interval $[E - \varepsilon, E + \varepsilon]$ and whose support is contained in $[E - 2\varepsilon, E + 2\varepsilon]$. Then the operator $\chi(H^W)$ defined by functional calculus satisfies the equation

$$\chi\left(H^W\right) \psi_{\hbar_m} = \psi_{\hbar_m}.$$

On the other hand we already stated that the symbol $\sigma\left(\chi(H^W)\right)$ is supported within $\mathrm{supp}(\chi(H_{\mathrm{cl}})) \subset H_{\mathrm{cl}}^{-1}([E - 2\varepsilon, E + 2\varepsilon])$. Thus for a phase space point $(\mathbf{p}, \mathbf{q})$ outside that thickened energy shell

$$\left(\varphi_{\mathbf{p},\mathbf{q}}, \psi_{\hbar_m}\right) = \left(\varphi_{\mathbf{p},\mathbf{q}}, \chi\left(H^W\right)\psi_{\hbar_m}\right) = \left(\chi\left(H^W\right)\varphi_{\mathbf{p},\mathbf{q}}, \psi_{\hbar_m}\right) = \mathcal{O}(\hbar^\infty),$$

so that $(\mathbf{p}, \mathbf{q})$ does not belong to the frequency set.

We considered an explicit example of that statement when we calculated in Subsect. 4.1 the Husimi function of the harmonic oscillator. In that case we saw for $n = 1$ spatial dimensions that the frequency set equalled the energy shell. However, for $n \geq 2$ dimensions the frequency set had the form of an at most n-dimensional submanifold of the $2n - 1 > n$-dimensional energy shell. Thus in that classically integrable case eigenfunctions are semiclassically much more localized than needed to meet the above relation (4.15).

If – as in the case of the harmonic oscillator – the quantum mechanical system is integrable, too (in the sense that the eigenfunctions of the Hamiltonians are the common eigenfunctions of n commuting $\hbar$-pseudodifferential operators), then one can understand such a behaviour, since then one expects semiclassical concentration on the common level set of all the n (independent) principal symbols.

We are interested here in quantum chaos, that is, classically ergodic motion, which is the extreme of classical behaviour opposite to integrability[5]. In that case we expect the eigenfunctions to be semiclassically *delocalized* on their energy shell, since there is no true invariant subset of positive measure which could support them. This is actually the content of the *Schnirelman Theorem* which we will analyse in some detail.

Clearly if we want to prove such a statement, it does not suffice to consider the *Hamiltonian function* which is the principal symbol of our operator, but we must consider the solution of the *Hamiltonian equations*, that is, dynamics.

---

[5]But these extremes coincide for $n = 1$ freedom, if the regular energy shell is compact and connected (and thus diffeomorphic to the circle S^1)

4.3 Fourier Integral Operators

Whereas a pseudodifferential operator is the quantum analog of multiplication by a phase space function, roughly speaking a Fourier integral operator (FIO) is the quantum analog of a canonical transformation[6].

For all times t the flow $\Phi^t : P \to P$ is such a canonical transformation, that is, a phase space diffeomorphism which leaves the symplectic two-form

$$\omega = \sum_{j=1}^{n} dq_j \wedge dp_j$$

invariant. Thus the unitary time evolution $U(t) = \exp(-iH_\hbar t/\hbar)$, which is the quantum analog of Φ^t, is a candidate for a Fourier integral operator.

The graph $\Lambda \subset P \times P$ of a canonical transformation $\Phi : P \to P$ is an $2n$-dimensional submanifold. It is a Lagrange submanifold of $P \times P$ w.r.t. the symplectic two-form

$$\Omega := \pi_I^* \omega - \pi_{II}^* \omega = \sum_{j=1}^{n} \left(dQ_j^I \wedge dP_j^I - dQ_j^{II} \wedge dP_j^{II} \right),$$

$\pi_k : P \times P \to P$ being the projections onto the factors. One can describe (at least locally) canonical transformations by generating functions.

Let us assume that Φ is a canonical transformation which is near to the identity in the sense that the tangent planes to the Lagrangean manifold Λ are uniformly near to the diagonal.[7]

Then we use the fact that the symplectic one-form

$$\Theta := \sum_{j=1}^{n} \left(Q_j^I dP_j^I + P_j^{II} dQ_j^{II} \right)$$

is closed on Λ (since $d\Theta = \Omega$ and $\Omega|_\Lambda = 0$).

On Λ which, being the graph of Φ, is diffeomorphic to $P \cong \mathbb{R}^{2n}$, we can integrate this one-form to obtain a function $S : \Lambda \to \mathbb{R}$ with $dS = \Theta|_\Lambda$ called the *generating function*. Since we assumed Φ to be near the identity, we can view S as a function

$$S \equiv S(\mathbf{P}^I, \mathbf{Q}^{II})$$

of the momenta P^I in the domain and the positions $\mathbf{Q}^{II}$ in the image. Comparing dS with Θ one deduces that

$$\mathbf{P}^{II} = \frac{\partial S}{\partial \mathbf{Q}^{II}} \quad \text{and} \quad \mathbf{Q}^I = \frac{\partial S}{\partial \mathbf{P}^I}.$$

---

[6]More precisely *unitary* FIO's can be interpreted in that way, whereas a general FIO is the quantum analog of a canonical transformation combined with multiplication by a phase space function

[7]This is the case if Φ is the Hamiltonian flow Φ^t for small $|t|$ of a function in $S^2(P)$.

We now introduce *Fourier Integral Operators* (FIO) U which correspond to the canonical transformation Φ and act on $\psi \in \mathcal{H}$ as

$$(U\psi)(\mathbf{Q}^{II}) := (2\pi\hbar)^{-n} \int_P \exp\left(i[S(\mathbf{P}^I, \mathbf{Q}^{II}) - \mathbf{P}^I \cdot \mathbf{Q}^I]/\hbar\right) \qquad (4.16)$$

$$\cdot \left(\sum_{j=0}^{\infty} a_j(\mathbf{P}^I, \mathbf{Q}^{II})\hbar^j\right) \psi(\mathbf{Q}^I) d\mathbf{P}^I d\mathbf{Q}^I,$$

with $a_j \in S^0(P)$ (or differ from such an operator by $\mathcal{O}(\hbar^\infty)$).

In the simplest case $\Phi = \text{Id}$ we have $S = \mathbf{P}^I \cdot \mathbf{Q}^{II}$ (up to an unimportant additive constant), so that U is in fact a pseudodifferential operator.

In order to understand how FIO's act on wave functions, we first ask how they move the frequency set. For simplicity we choose for the wave function ψ in (4.17) the coherent state $\varphi_{\mathbf{p}^I, \mathbf{q}^I}$ centered at $(\mathbf{p}^I, \mathbf{q}^I) \in P$, see (4.5). Remember that in order to find the frequency set of $U\varphi_{\mathbf{p}^I, \mathbf{q}^I} \in \mathcal{H}$ we must consider its Husimi function which means estimating

$$\left(\varphi_{\mathbf{p}^{II}, \mathbf{q}^{II}}, U\varphi_{\mathbf{p}^I, \mathbf{q}^I}\right) \qquad \text{for} \quad (\mathbf{p}^{II}, \mathbf{q}^{II}) \in P.$$

But[8] this equals the lengthy expression

$$(\pi\hbar)^{-n/2} \int_{\mathbb{R}_\mathbf{q}^n} (2\pi\hbar)^{-n} \int_P \exp\left(i[S(\mathbf{P}^I, \mathbf{Q}^{II}) + (\mathbf{p}^I - \mathbf{P}^I)\mathbf{Q}^I - \mathbf{p}^{II}\mathbf{Q}^{II}]/\hbar\right)$$

$$e^{-((\mathbf{Q}^I - \mathbf{q}^I)^2 + (\mathbf{Q}^{II} - \mathbf{q}^{II})^2)/2\hbar} \left(\sum_j a_j(\mathbf{P}^I, \mathbf{Q}^{II})\hbar^j\right) d\mathbf{Q}^{II} \, d\mathbf{P}^I d\mathbf{Q}^I.$$

- Integrating out $\mathbf{P}^I$, the non-stationary phase formula tells us that we get the leading contribution in $\hbar$ if $\frac{\partial}{\partial \mathbf{P}^I}[\cdots] = 0$, that is,

$$\mathbf{Q}^I = \frac{\partial S(\mathbf{P}^I, \mathbf{Q}^{II})}{\partial \mathbf{P}^I}.$$

- Integrating out $\mathbf{Q}^{II}$, we obtain the conditions

$$\mathbf{P}^{II} = \frac{\partial S(\mathbf{P}^I, \mathbf{Q}^{II})}{\partial \mathbf{Q}^{II}} \quad \text{and} \quad \mathbf{q}^{II} = \mathbf{Q}^{II}$$

- Similarly, integrating over $\mathbf{Q}^I$ gives

$$\mathbf{q}^I = \mathbf{Q}^I$$

---

[8] up to the phase $\exp(i(\mathbf{p}^{II}\mathbf{q}^{II} - \mathbf{p}^I\mathbf{q}^I)/\hbar)$

viz. the frequency set

$$\mathrm{FS}\left(U\varphi_{\mathbf{p}^I,\mathbf{q}^I}\right) \subset \{\Phi((\mathbf{p}^I,\mathbf{q}^I))\} = \Phi\left(\mathrm{FS}(\varphi_{\mathbf{p}^I,\mathbf{q}^I})\right).$$

One may conclude that the frequency set of a general wave function is transported by a FIO with its canonical transformation.

Our original motivation for introducing FIO's was to consider quantum dynamics, that is, the unitary group

$$U(t) = \exp(-iH^{\mathrm{W}}t/\hbar) : \mathcal{H} \to \mathcal{H}. \tag{4.17}$$

In that case the generating function S is time-dependent and satisfies the so-called *Hamilton-Jacobi* equation

$$\begin{cases} \partial_t S(t,\mathbf{P}^I,\mathbf{Q}^{II}) &= -H\left(\partial_{\mathbf{Q}^{II}} S(t,\mathbf{P}^I,\mathbf{Q}^{II}),\mathbf{Q}^{II}\right) \\ S(0,\mathbf{P}^I,\mathbf{Q}^{II}) &= \mathbf{P}^I \cdot \mathbf{Q}^{II} \end{cases}$$

which is a first order partial differential equation (P.D.E.).

Example. For the harmonic oscillator $H(p,q) = \frac{1}{2}(p^2 + q^2)$ we obtain

$$S(t,\mathbf{P}^I,\mathbf{Q}^{II}) = \frac{\mathbf{P}^I\mathbf{Q}^{II} - \frac{1}{2}((\mathbf{P}^I)^2 + (\mathbf{Q}^{II})^2)\sin t}{\cos t}, \qquad |t| < \pi/2.$$

By the Hamilton-Jacobi equation the Schrödinger equation

$$\left(i\hbar\frac{d}{dt} - H^{\mathrm{W}}\right)U(t) = 0, \qquad U(0) = \hat{\mathbb{1}}$$

is met by the family $U(t)$ of FIO's with phase function S to leading order if $a_0(t = 0) = 1$ and $a_j(t = 0) = 0$ for $j \geq 1$.

For the time-dependent coefficients a_j of the FIO we obtain so-called *transport equations* which are first order P.D.E.

Actually we are interested only to describe time evolution in a bounded energy range and thus may try to approximate $U(t) \cdot \chi(H^{\mathrm{W}})$. This is actually possible for some small time interval $|t| < T$ up to order $\mathcal{O}(\hbar^\infty)$. Then by the group property $U(t_1 + t_2) = U(t_1) + U(t_2)$ one extends this approximation to any bounded interval.

- So in the *Schrödinger picture* of quantum mechanics we can study semiclassically the time evolution

$$t \mapsto \psi_t := U(t)\psi$$

of a wave function $\psi \in \mathcal{H}$ and the expectations $(\psi_t, O^{\mathrm{W}}\psi_t)$

- Alternatively in the *Heisenberg picture* we leave the wave function invariant
  and evolve the operators:

$$t \mapsto (O^{\mathrm{W}})_t := U(-t)O^{\mathrm{W}}U(t).$$

Here $(O^{\mathrm{W}})_t$ is the composition of three Fourier integral operators. The canonical transformation connected with $U(-t)$ is the inverse of the one of $U(t)$, and the canonical transformation of O is the identity, since O^{W} is a pseudodifferential operator. Thus the canonical transformation of the composed operator, being the composition of the three transformations, is the identity. It even turns out that $(O^{\mathrm{W}})_t$ is a pseudodifferential operator, too.

If we assume the symbol O to be $\in S_\hbar^0(P)$ (and to be of compact support), then we have

Theorem 4.3 (Egorov)

$$\|(O^{\mathrm{W}})_t - (O \circ \Phi^t)^{\mathrm{W}}\| = \mathcal{O}(\hbar)$$

uniformly in any bounded interval $|t| < T$. That means that for bounded time the quantum time evolution follows the classical flow Φ^t up to an error which becomes small in the semiclassical limit.

4.4   The Schnirelman Theorem

We know that independent of whether the integrable is integrable or not, the semiclassical eigenfunctions are microlocalized in their energy shell $H_{\mathrm{cl}}^{-1}(E) \subset P$.

If $F \in S_\hbar^0(P)$ and all the derivatives have finite L^1-norm, then $F^{\mathrm{W}} \in B(\mathcal{H})$ is a trace class operator with trace

$$\mathrm{tr}(F^{\mathrm{W}}) = (2\pi\hbar)^{-n} \int_P F(\mathbf{p}, \mathbf{q}) \, d\mathbf{p}d\mathbf{q} \tag{4.18}$$

(Theorem II-53 of [16]). This is a consequence of the fact that the trace is the space integral

$$\mathrm{tr}(F^{\mathrm{W}}) = \int_{\mathbb{R}_\mathbf{q}^n} K_F(\mathbf{q}, \mathbf{q}) d\mathbf{q}$$

of the kernel K_F of F^{W}, and the defining formula (4.10) for Weyl quantization.

From this we can infer that under our standard assumptions on the Hamiltonian H_{cl} the number of eigenvalues in an interval $[E_1, E_2]$ is asymptotic to

$$(2\pi\hbar)^{-n} \int_P \chi_{[E_1,E_2]}(H_{\mathrm{cl}}(\mathbf{p}, \mathbf{q})) \, d\mathbf{p}d\mathbf{q}, \tag{4.19}$$

$\chi_{[E_1,E_2]}$ being the characteristic function of the interval. This *Weyl term* conforms with the intuition that each quantum state covers a volume of $(2\pi\hbar)^n$ in phase space.

In the proof one first shows that for a *smooth* cutoff function $\chi \in C_0^\infty(\mathbb{R})$ with $\chi(E) = 1$ for $E_1 \leq E \leq E_2$

$$\text{tr}\left(\chi(H_{\text{cl}}^{\text{W}})\right) = (2\pi\hbar)^{-n}\left(\int_P \chi(H_{\text{cl}}(\mathbf{p}, \mathbf{q}))\, d\mathbf{p}d\mathbf{q} + \mathcal{O}(\hbar)\right),$$

using formula (4.18) for the trace.

But then the hard work consists in going from the smooth function χ to the characteristic function $\chi_{[E_1, E_2]}$.

Our initial example of the n-dimensional harmonic oscillator tells us why: If all its frequencies are equal: $\omega_j = \omega$, then the eigenvalue $\hbar\omega(\frac{n}{2} + k)$ has multiplicity $\binom{k+n-1}{k} \sim \frac{k^{n-1}}{(n-1)!}$. This large degeneracy implies – since for a given energy the integer k scales like $\hbar^{-1}$ – that the *fluctuations* in the number $\text{tr}(\chi_{[E_1, E_2]}(H_{\text{cl}}^{\text{W}}))$ of eigenvalues are of order $\hbar^{1-n}$, that is, next to leading order.

On the other hand, if the frequencies of the harmonic oscillator are *not* rationally related, there are no degeneracies, and the fluctuations are smaller.

Generalizing the example of the harmonic oscillator, in general the fluctuations are small if the closed orbits on the energy shell have measure zero: Under that condition[9] one may consider the eigenfunctions $\psi_{\hbar, j}$ with $H_{\text{cl}}^{\text{W}}\psi_{\hbar, j} = E_{\hbar, j}\psi_{\hbar, j}$ and their Husimi functions $\rho_{\hbar, j}$ in an small energy window $[E - c\hbar, E + c\hbar]$. The cardinality of the index set

$$\Lambda_\hbar := \{j \in \mathbb{N} \mid |E_{\hbar, j} - E| \leq c\hbar\}$$

then diverges in the semiclassical limit.

Theorem 4.4 ([10]) *The Husimi functions are converging weakly* **in the mean** *to the Liouville measure λ_E:*

$$\lim_{\hbar \searrow 0} \frac{1}{|\Lambda_\hbar|} \sum_{j \in \Lambda_\hbar} \rho_{\hbar, j} = \lambda_E. \tag{4.20}$$

But as we saw in the example of the $n \geq 2$-dimensional harmonic oscillator, in general the *individual* eigenfunctions are *not delocalized* on the whole energy shell and can converge to lower-dimensional subsets.

However this delocalization occurs in the quantum-chaotic situation. This is the context of the *Schnirelman Theorem* [18], which was proved in different contexts by Zelditch [21], Colin de Verdiere [5], Helffer, Martinez and Robert [10], and others. Here the version of [10]:

Theorem 4.5 *If the classical flow Φ^t is ergodic on Σ_E, then for all symbols $O \in S_\hbar^0(P)$ and $\varepsilon > 0$*

$$\lim_{\hbar \searrow 0} \frac{\left|\left\{j \in \Lambda_\hbar \,\middle|\, |(\psi_{\hbar, j}, O^{\text{AW}}\psi_{\hbar, j}) - \int_{\Sigma_E} O d\lambda_E| < \varepsilon\right\}\right|}{|\Lambda_\hbar|} = 1.$$

---

[9]which implies that $n \geq 2$

In words, up to a subset of wave functions of density zero all quantum expectations approach the classical expectations.

Proof. First we localize in energy by substituting $H^W \cdot \chi(H^W)$ for H^W. This does neither change the eigenfunctions $\psi_{\hbar,j}$, $j \in \Lambda_\hbar$ nor the flow Φ^t near Σ_E.

Basically we would like to argue as follows: The above difference is small since

1. by ergodicity the *space average*

$$\bar{O} := \int_{\Sigma_E} O d\lambda_E$$

is equal to the *time average*

$$\lim_{T \to \infty} \frac{1}{T} \int_0^T O(\Phi^t(x)) \, dt, \qquad x \in \Sigma_E,$$

and since

2.

$$
\begin{aligned}
(\psi_{\hbar,j}, O^{AW} \psi_{\hbar,j}) &= \frac{1}{T} \int_0^T (\psi_{\hbar,j}, O^{AW} \psi_{\hbar,j}) dt \\
&= \int_P \left(\frac{1}{T} \int_0^T O(\Phi^t(x)) dt \right) d\rho_{\hbar,j}(x) + \mathcal{O}(\hbar),
\end{aligned}
$$

using Egorov's theorem and

$$\|F^{AW} - F^W\|_{B(\mathcal{H})} = \mathcal{O}(\hbar)$$

(formula (4.13)).

The 'ergodic' error term

$$E_{\hbar,j}^T := \int_P \left| \frac{1}{T} \int_0^T O(\Phi^t(x)) dt - \bar{O} \right| d\rho_{\hbar,j}(x)$$

in 1) should become small if $T \to \infty$, and for given time T, the '**quantization**' error term

$$Q_{\hbar,j}^T := \left| \int_P \left(O(x) - \frac{1}{T} \int_0^T O(\Phi^t(x)) dt \right) d\rho_{\hbar,j}(x) \right|$$

vanishes in the semiclassical limit $\lim_{\hbar \searrow 0}$.

However there is a flaw in this argument. In general the time average of ergodic flows does *not* converge pointwise to the space average, but only for a set of full measure. If $\rho_{\hbar,j}$ concentrates semiclassically on exceptional points $x \in \Sigma_E$, the difference does not vanish.

However, these Husimi functions are then themselves exceptional. Namely the 'bad set'

$$B \equiv B(T,\varepsilon) := \left\{ x \in \Sigma_E \left| \left| \frac{1}{T} \int_0^T O(\Phi^t(x))\, dt - \bar{O} \right| \geq \frac{\varepsilon}{4} \right\} \right.$$

can be made smaller than

$$\lambda_E(B(T,\varepsilon)) \leq \frac{\delta}{2}$$

in measure by averaging up to time $T \geq T(\delta)$.

By making $\hbar \leq \hbar(\delta)$ we can by the convergence in the mean result (4.20) ensure that

$$\left(\frac{1}{|\Lambda_\hbar|} \sum_{j \in \Lambda_\hbar} \rho_{\hbar,j} \right) B(T,\varepsilon)) \leq \delta.$$

Then by the Tchebychev inequality

$$\left| \left\{ j \in \Lambda_\hbar \,\middle|\, \rho_{\hbar,j}(B) \geq \sqrt{\delta} \right\} \right| \leq \sqrt{\delta} |\Lambda_\hbar|$$

so that for the indices j in the complementary subset of $\Lambda_\hbar$

$$E_{\hbar,j}^T \leq \frac{\varepsilon}{4} \int_{\Sigma_E - B} \lambda_E + 2\sqrt{\delta} \|O^W\| \leq \frac{\varepsilon}{2}$$

for $\delta \leq \delta_0$. In fact for all $0 < \delta \leq (\varepsilon/8 \sup_\hbar \|O^W\|)^2$ there exists $\hbar_{\max}(\delta)$ with

$$Q_{\hbar,j}^T < \frac{\varepsilon}{2} \qquad \text{for } \hbar \leq \hbar_{\max}(\delta).$$

Thus in the semiclassical limit $\hbar \searrow 0$ we can move δ to zero so that we have proven the theorem. □

Remarks 4.6 • *One can even prove that the individual Husimi functions $\rho_{\hbar,j}$ converge weakly to the Liouville measure λ_E if one restricts oneself to a subset $j \in \Lambda_\hbar$ of density one [10]. So most eigenfunctions are semiclassically delocalized by quantum chaos.*

For negatively curved Riemannian manifolds Zelditch [22] proved logarithmic upper bounds for the mean approach to the Liouville measure.

• *As an application it was shown in [13] that the group velocity of an electron in a planar crystal with attracting Coulombic potential vanishes in the semiclassical limit. Thus the quantum particle is ballistic but slowed down since it remembers that the classical motion is a deterministic diffusion process [12].*

- *All weak-* limit points of Husimi function measures must be flow-invariant measures, like λ_E. However, in general there exist lots of them, even ergodic ones. In particular each measure equidistributed on a closed orbit is ergodic, and there has been some numerical evidence for a semiclassical concentration near such unstable orbits ('scars').*

On the other hand, Luo and Sarnak [14] proved a kind of 'unique quantum ergodicity' result for the continuous spectral subspace of the Laplacian on the modular domain, implying that in that case the Liouville measure λ_E is the only limit measure.

4.5 Further Directions

> "... the above statements are false in various
> special cases (more or less any time when we can
> completely analyze the spectrum!) so that they are
> at best true in some generic sense." P. Sarnak [17]

Our basic definition of quantum chaos as the quantum mechanics of a classically ergodic system is not very satisfying from a conceptual point of view. One would like to find an intrinsic, that is, quantum mechanical, criterion for quantum chaos. In fact a large proportion of the physics-oriented literature is devoted to that question.

However in this respect Gutzwiller states in [8]:

> "Quantum mechanics has liberated us from the scourge of classical chaos, and we will find that the symptoms of chaos are hard to pin down in this new environment."

In fact quantum mechanics is much less chaotic than classical mechanics.

This statement does not seem to conform with the notion of quantum mechanics as a theory intrinsically connected with uncertainty. But one has to keep in mind that here we are not dealing with information loss connected with measurement but with information loss caused by the dynamics (to be sure, the two aspects are sometimes hard to disentangle).

Maybe the easiest way to see why quantum mechanics is less chaotic is by the example of the motion on a closed (i.e., compact and without boundary) Riemannian manifold M of negative sectional curvature.

Here the 'classical' geodesic motion on the unit tangent bundle SM of M is known to be have the best ergodic properties. In particular any probability measure on SM which is absolutely continuous w.r.t. Liouville measure converges to Liouville measure in the time $t \to \infty$ limit.

The 'quantum' motion is generated by the Laplace-Beltrami operator which has purely discrete spectrum $E_1 = 0 < E_2 \leq E_3 \leq \ldots E_j \leq \ldots$, with eigenfunctions $\psi_j \in L^2(M)$. Thus any normalized wave function $\varphi \in L^2(M)$ can be written in the form $\varphi = \sum_{j=1}^{\infty} c_j \psi_j$ with $\sum_{j=1}^{\infty} |c_j|^2 = 1$. At time t

$$\varphi(t) = \sum_{j=1}^{\infty} c_j e^{-iE_j t} \psi_j.$$

Now there exists a sequence of times t_k with $\lim_{k \to \infty} t_k = \infty$ and

$$\lim_{k \to \infty} \varphi(t_k) = \varphi,$$

that is, the wave function returns again and again in any small neighbourhood of its initial state. This can be first proven for finitely many non-zero coefficients c_j and then for arbitrary $l^2(\mathbb{N})$ sequences of coefficients by an approximation argument.

In particular this implies that the Husimi function of φ does not equidistribute in the $t \to \infty$ limit if it does not already have this property for $t = 0$.

Another way to see the absence of memory loss in the quantum mechanics of finitely many particles is to consider the *quantum dynamical entropy* by Connes, Narnhofer and Thirring [6], as discussed in the book [2] by Benatti. This is a quantum generalization of the classical Kolmogorov-Sinai entropy. However, whereas in the above example the geodesic flow has positive entropy, the quantum motion has zero entropy.

In fact as we already mentioned one can only hope to detect quantum chaos in the semiclassical limit (which in the example of the Laplace-Beltrami operator is equivalent to the high energy limit).

Thus it is natural to ask whether the Schnirelman theorem on the eigenfunctions has a counterpart concerning the eigenvalues of a quantum chaotic Hamiltonian.

In (4.19) we stated that the number of eigenvalues of a pseudodifferential Hamiltonian in some energy interval is in leading order given by the phase space volume of that energy range, divided by the volume element $(2\pi\hbar)^n$. This *Weyl asymptotics* is the same for classically integrable resp. ergodic motion.

So searching for a fingerprint of quantum chaos in the eigenvalue distribution, one has to consider its fluctuations around the Weyl term.

One proposed criterion is the statistics of nearest neighbour spacing $E_{j+1} - E_j$ in some small energy interval as $\hbar \searrow 0$. After division through the above mean Weyl density many numerical experiments show the following dichotomy

- For classically *integrable* systems the level spacing probability distribution P is of the Poisson form $P(s) = e^{-s}$.

- In the *chaotic* case one has *level repulsion* with $P(s) \sim c \cdot s$ for $s \searrow 0$.

Analytical arguments supporting this picture are given by Berry in [3].

It is clear from the outset that some additional assumptions are necessary to ensure the mere existence of a limiting distribution.

In the integrable cases of a harmonic oscillator with rationally related frequencies or motion on a perfect sphere the large eigenvalue degeneracies give rise to a level spacing concentrated at 0.

Even worse, on the ergodic side the Laplace-Beltrami operator on certain (arithmetic) surfaces of constant negative curvature like the modular domain does not show the expected behaviour. Instead a Poissonian distribution was found by Bolte et al. [4].

However, under suitable genericity assumptions there seems to be some truth in the above statement. In this respect consider Chapter 7 by Ya. Sinai on Liouville surfaces and the statements in [17] on flat tori.

Generally speaking, like in the above analysis of the Schnirelman Theorem, mathematical research on quantum chaos is based on semiclassical calculus, whereas in the physics literature is based on *Gutzwiller's formula* [8]. This is a formula relating classical periodic orbits and the quantal spectrum and may be thought of as a generalization of the Selberg trace formula.

The basic idea is to use the unstable closed orbits of a classically chaotic system in a way which is similar to the Bohr-Sommerfeld quantization based on the invariant tori of integrable systems. The technical problem of that approach consists in the exponential proliferation of the number of these orbits in their length.

Bibliography

[1] Abraham, R., Marsden, J.E.: Foundations of Mechanics. Reading: Benjamin/Cummings; 2.Ed. 1978

[2] Benatti, F.: Deterministic Chaos in Infinite Quantum Systems. Trieste Notes in Physics. Berlin: Springer 1993

[3] Berry, M.: Semiclassical Mechanics of Regular and Irregular Motion. In: Comportement chaotique des systèmes déterministes. Les Houches XXXVI. Amsterdam: North-Holland 1981

[4] Bolte, J., Steil, G., Steiner, F.: Arithmetical Chaos and Violation of Universality in Energy Level Statistics. Phys. Rev. Lett. **69**, 2188–2191 (1992)

[5] Colin de Verdiere: Ergodicité et fonctions propres du Laplacien. Commun. Math. Phys. **102**, 497–502 (1985)

[6] A. Connes, H. Narnhofer & W. Thirring: *Dynamical Entropy for C^* algebras and von Neumann Algebras, Commun. Math. Phys.* **112**, 691 (1987)

[7] Fefferman, C.: The Uncertainty Principle. Bull. AMS **9**, 129–206 (1983)

[8] Gutzwiller, M.: Chaos in Classical and Quantum Mechanics. Berlin, Heidelberg, New York: Springer; 1990

[9] Helffer, B.: h-pseudodifferential operators and applications: an introduction. Preprint 1995

[10] Helffer, B., Martinez, A., Robert, D.: Ergodicité et limite semi-classique. Commun. Math. Phys. **109**, 313–326 (1987)

[11] Hörmander, L.: The Analysis of Linear Partial Differential Operators. Vols. I–IV. Grundlehren der mathematischen Wissenschaften 256, 257, 274, 275 Berlin: Springer 1983–1985

[12] Knauf, A.: Ergodic and Topological Properties of Coulombic Periodic Potentials. Commun. Math. Phys. **110**, 89–112 (1987)

[13] Knauf, A.: Coulombic Periodic Potentials: The Quantum Case. Annals of Physics **191**, 205–240 (1989)

[14] Luo, W., Sarnak, P.: Number Variance for Arithmetic Hyperbolic Surfaces. Comm. Math. Phys. **161**, 419–431 (1994)

[15] Reed, M., Simon, B.: Methods of Modern Mathematical Physics. Vol. I: Functional Analysis. Vol. II: Fourier Analysis, Self Adjointness. Vol. IV: Analysis of Operators. New York: Academic Press 1972, 1975, 1978

[16] Robert, D.: Autour de l'approximation semi-classique. Boston, Basel, Stuttgart: Birkhäuser 1987

[17] Sarnak, P.: Spectra and Eigenfunctions of Laplacians. Lecture Notes 1995

[18] Schnirelman, A.I.: Ergodic Properties of Eigenfunctions. Usp. Math. Nauk. **29**, 181–182 (1974)

[19] Simon, B.: The Classical Limit of Quantum Partition Functions. Commun. Math. Phys. **71**, 247–276, (1980)

[20] Thirring, W.: Lehrbuch der Mathematischen Physik, Band 3. Springer 1979

[21] Zelditch, S.: Uniform Distribution of Eigenfunctions on Compact Hyperbolic Surfaces. Duke Math. J. **55**, 919–941 (1987)

[22] Zelditch, S.: On the Rate of Quantum Ergodicity I: Upper Bounds. Commun. Math. Phys. **160**, 81–92 (1994)

Chapter 5

Ergodicity and Mixing

The notion of ergodicity was introduced by L. Boltzmann in connection with Foundations of Statistical Mechanics. Now its role for Statistical Mechanics is not so much clear but it is very important for the theory of dynamical systems and deterministic chaos.

The main definition is given in the framework of one-parameter groups or semi-groups of measure-preserving transformations.

Let $(M, \mathcal{M}, \mu)$ be a measure space with probability measure μ. Consider a one-parameter group or semi-group of measure-preserving transformations $\{T^t\}$.

Theorem 5.1 (Birkhoff Ergodic Theorem.) *For every*
$f \in L^1(M, \mathcal{M}, \mu)$ *and a.e. $x \in M$ there exist the time-averages*

1. *in the case of semi-groups*

$$\lim_{t \to \infty} \frac{1}{t} \sum_{k=0}^{t-1} f(T^k x) = \bar{f}^+(x) \quad \text{(discrete time)}$$

$$\lim_{t \to \infty} \frac{1}{t} \int_0^t f(T^s x) ds = \bar{f}^+(x) \quad \text{(continuous time)}$$

2. *in the case of groups*

$$\lim_{t \to \infty} \frac{1}{t} \sum_{k=0}^{t-1} f(T^{\pm k} x) = \bar{f}^\pm(x) \quad \text{(discrete time)}$$

$$\lim_{t \to \infty} \frac{1}{2t+1} \sum_{k=-t}^{t} f(T^{\pm k} x) = \bar{f}(x) \quad \text{(discrete time)}$$

$$\lim_{t \to \infty} \frac{1}{t} \int_0^t f(T^{\pm s} x) ds = \bar{f}^\pm(x) \quad \text{(continuous time)}$$

$$\lim_{t \to \infty} \frac{1}{2t} \int_{-t}^{t} f(T^s x) ds = \bar{f}(x) \quad \text{(continuous time)}.$$

In the case of groups $\bar{f}^+(x) = \bar{f}(x) = \bar{f}^-(x)$ a.e. The limiting function $\bar{f}^+(x)$ is invariant (mod 0), $\bar{f}^+(T^s x) = \bar{f}^+(x)$ a.e., and

$$\int_M \bar{f}^+(x)d\mu(x) = \int_M \bar{f}(x)d\mu(x).$$

Definition 5.2 *A dynamical system is called* ergodic *if every invariant (mod 0) function h ($h(T^s x) = h(x)$ for a.e. x; the set of x where the last equality holds may depend on s) is constant, i.e. h =const a.e.*

In view of the Birkhoff ergodic theorem ergodicity means that $\bar{f}^+(x) = \int f(z)d\mu(z)$ a.e., i.e. time averages are equal to space averages.

Decomposition of an arbitrary invariant measure into ergodic components: Under very mild conditions on the measure space one can show (von Neumann, Rokhlin) that a dynamical system can be modified on a subset of measure zero so that for the new system the space M has a measurable partition $\zeta^{(\text{inv})}$ such that each element $C_{\zeta^{(\text{inv})}}$ is invariant under the action of the modified system. The restriction to $C_{\zeta^{(\text{inv})}}$ preserves the conditional measure and is ergodic. The representation $d\mu = d\mu(C_{\zeta^{(\text{inv})}})d\mu(x \mid C_{\zeta^{(\text{inv})}})$ where $d\mu(C_{\zeta^{(\text{inv})}})$ is the induced measure on the factor space $M|_{\zeta^{(\text{inv})}}$ is called *desintegration* of measure or *decomposition* of μ into ergodic components.

The main problem of ergodic theory is to construct invariant measures for given dynamical systems and to find their decompositions into ergodic components.

Remarks 5.3   *1. A dynamical system is called* uniquely ergodic *if it has only one invariant probability measure. This measure is certainly ergodic. If M is a compact topological space and $\mathcal{M}$ is its Borel σ-algebra, then unique ergodicity can only happen if for any continuous function f time averages exist everywhere and are constants.*

Known examples of uniquely ergodic dynamical systems include shifts on compact abelian groups (Weyl) and skew rotations on tori (Furstenberg). For interval exchange transformations the number of invariant measures does not exceed the number of exchanged intervals (generically). This was proven by Oseledets. Veech used the methods of algebraic geometry to develop a deep theory of interval exchange transformations.

2. The problem of existence of an invariant measure for a given action $\{T^t\}$ has a positive answer in the case of continuous one-parameter groups or semi-groups of transformations of compact topological spaces (Bogoliubov-Krylov theory).

However, in the case of smooth manifolds and smooth actions it has a different aspect. Take any absolutely continuous measure μ_0 and its shift μ_t, i.e. $\mu_t(C) = \mu_0((T^t)^{-1}C)$, $C \in \mathcal{M}$. It is natural to ask under which conditions μ_t converges to a limit μ not depending on μ_0. This phenomenon is a manifestation of irreversibility in deterministic dynamics. At present some results about such type of behavior are proven under some conditions of instability (hyperbolicity).

3. There are various extensions of Birkhoff's Theorem to actions of other groups of transformations (multi-parameter abelian groups, nilpotent groups). However, the direct generalization to groups close to free groups is wrong. Time averages can, if they exist, depend on the form of the sets over which one averages given functions.

4. There are many problems in harmonic analysis which are close to the problem of existence of the limit of time averages. To give an example, consider a periodic function $h(x_1, \ldots, x_r)$ of r variables. Fix a point $x^{(0)} = (x_1^{(0)}, \ldots, x_r^{(0)})$ and denote by $S_r(x^{(0)})$ the sphere of radius r whose center is $x^{(0)}$. We can consider the average $A_r(x^{(0)})$ of h over $S_r(x^{(0)})$. The convergence of $A_r(x^{(0)})$ to $\int h(x_1, \ldots, x_r) dx_1, \ldots, dx_r$ everywhere is a simple statement in the case of continuous functions h. However, if $h \in L^p$ then the problem becomes difficult and the answer depends on p (Bourgain).

5. In Hamiltonian systems one has a natural invariant measure. Let M_0 be a phase space of a Hamiltonian system with r degrees of freedom. Then $\dim(M_0) = 2r$ and if ω is a nondegenerate symplectic form we can choose canonical coordinates $\mathbf{p}, \mathbf{q}$ so that in the autonomous case the equations of motion take the form

$$\frac{dq_i}{dt} = \frac{\partial H}{\partial p_i}, \qquad \frac{dp_i}{dt} = -\frac{\partial H}{\partial q_i}. \tag{5.1}$$

Here $H(\mathbf{p}, \mathbf{q})$ is a Hamiltonian about which we assume that it does not depend on time. Then H is a first integral and in many cases the manifold $\{H(\mathbf{p}, \mathbf{q}) = \text{const}\} = M$ is compact. The system (5.1) preserves the Liouville measure $\prod_{i=1}^{r} dp_i dq_i$. Its restriction to M generates a finite invariant measure which can be normalized. It is called in statistical mechanics the microcanonical distribution. The famous Ergodic Hypothesis is the problem about ergodicity of the dynamical system (5.1) restricted to M.

6. There is a big field of research now connected with ODE or PDE having random terms in the right hand side. A typical example is

$$\frac{dx_i}{dt} = f_i(x) + d\xi_i(x, t), \qquad 1 \le i \le n$$

where $\xi(x, t)$ is a stochastic flow. Probability distributions for solutions constitute Markov processes and any invariant measure for this process generates a dynamical system of shifts preserving the measure of the Markov process. Ergodicity of the Markov process implies ergodicity of the dynamical systems.

Definition 5.4 *A dynamical system* $(M, \mathcal{M}, \mu, \{T^t\})$ *is called* mixing
if for any $f, h \in L^2$

$$\lim_{t \to \infty} \int f(T^t x) h(x) d\mu(x) = \int f(x) d\mu(x) \cdot \int h(x) d\mu(x).$$

Mixing implies some irreversibility of the dynamics. Consider an absolutely con-
tinuous (with respect to μ) measure μ_0, $\frac{d\mu_0}{dx} = f(x)$. Then its shift μ_t, $\mu_t(C) = \mu_0((T^t)^{-1}C)$ is absolutely continuous with respect to μ and its density is $f(T^t x)$
$(\int h(x) d\mu_t(x) = \int h(x) f(T^t x) d\mu(x))$.
    In the case of mixing the integrals converge to $\int h(x) d\mu(x)$ provided
$\int f(x) d\mu(x) = 1$ but this does not mean $f(T^t x) \to 1$ point-wise.
    Usually the density $f(T^t x)$ becomes a very irregular function and the con-
vergence is only in the mean.
    I.R. Prigogine formulated a general problem to find a functional $\mathcal{H}(\mu_0)$ on
the space of absolutely continuous measures whose value grows with time, i.e.
$\mathcal{H}(\mu_t) \geq \mathcal{H}(\mu_0)$, $t > 0$ and $\max_{\mu_0} \mathcal{H}(\mu_0) = \mathcal{H}(\mu)$.
    Some candidates for $\mathcal{H}$ can be proposed for hyperbolic dynamical systems
and it is hard to imagine that there is a universal $\mathcal{H}$ for all mixing systems.
    There are several generalizations of the notion of mixing but we shall propose
only one.
    Take two sets $A, B \in \mathcal{M}$. Denote by $\mathcal{M}_t^\infty(A)$ the least σ-subalgebra contain-
ing all sets $T^s A$, $s \geq t$.

Definition 5.5 *A dynamical system* $(M, \mathcal{M}, \mu, \{T^t\})$ *is called* $\mathbb{K}$-system *if for any*
A the intersection

$$\bigcap_{t>0} \mathcal{M}_t^\infty(A) = \mathcal{M}_\infty^\infty(A) = \mathcal{N}$$

where $\mathcal{N}$ *is the trivial* σ-*algebra of subsets of probability 1 or 0, i.e. what is the*
same,

$$\mu(B \mid \mathcal{M}_t^\infty(A)) \to \mu(B) \qquad a.e. \ as \ t \to \infty.$$

Mixing systems are ergodic but there are ergodic systems which are not mixing
(group translations on abelian groups). $\mathbb{K}$-systems are mixing but there is a lot
of mixing systems which are not $\mathbb{K}$-systems. However in hyperbolic systems it is
easier to prove that a dynamical system is a $\mathbb{K}$-system and to derive ergodicity
and mixing as a corollary of this statement.
    A function $f(x)$ is called an eigen-function of the dynamical system if $f \in L^2$
and $f(T^t x) = e^{i\lambda t} f(x)$ a.e. for some constant λ.

Definition 5.6 *A dynamical system* $(M, \mathcal{M}, \mu, \{T^t\})$ *is called* weakly mixing *if each*
eigen-function is constant, i.e. $f = $ const *a.e.*

Weakly mixing dynamical systems are ergodic but there are weakly mixing systems
which are not mixing.

Chapter 6

Expanding Maps

We shall begin with some purely probabilistic statements. Consider the phase space of a symbolic Markov chain M. More precisely, let M_0 be the space of sequences $\omega = \{\ldots, \omega_{-n}, \ldots, \omega_0, \omega_1, \ldots, \omega_m, \ldots\}$ where each symbol ω_k takes one of r values i.e. $\omega_k \in \{1, \ldots, r\}$.

Let $A = \|a_{ij}\|$ be a matrix for which $a_{ij} = 1$ or 0. Denote by $M = M_0(A) \subset M_0$ the subset of those ω for which

$$a_{\omega_t \omega_{t+1}} = 1, \qquad -\infty < t < \infty.$$

We assume that A is transitive, i.e. for some k its kth power $A^k = \|a_{ij}^{(k)}\|$ consists of positive matrix elements.

Take a stochastic matrix $P = \|p_{ij}\|$ subordinated to A, i.e. $p_{ij} > 0$ iff $a_{ij} > 0$. The usual ergodic theorem for Markov chains says that there is a unique measure μ on the Borel σ-algebra of subsets of M_0 which is invariant under the shift, Markov and

$$\mu\left(\omega_s = j \mid \omega_{s-1} = i, \omega_{s-2} = i_{-2}, \omega_{s-3} = i_{-3}, \ldots\right) = p_{ij}$$

a.e. We shall consider a generalization of this theorem.

Assume that we are given a function

$$p(j \mid i_{-1}, i_{-2}, i_{-3}, \ldots, i_{-k}, \ldots), \qquad i, j \in \{1, \ldots, r\}$$

with the following properties:

i_1) $p(j \mid i_{-1}, i_{-2}, \ldots, i_{-k}, \ldots) = 0$ if $a_{i_{-1}j} = 0$;
$p(j \mid i_{-1}, i_{-2}, \ldots, i_{-k}, \ldots) \geq \text{const}$ if $a_{i_{-1}j} > 0$.

i_2) $\sum_j p(j \mid i_{-1}, i_{-2}, \ldots, i_{-k}, \ldots) = 1$.

i_3) If

$$\delta_k = \sup_{\substack{i_{-1}, \ldots, i_{-k} \; i'_{-s}, i''_{-s}, s \geq k+1 \\ j: p_{i_{-1}j} > 0}} \sup \left| \frac{p(j \mid i_{-1}, i_{-2}, \ldots, i_{-k}, i'_{-k-1}, \ldots)}{p(j \mid i_{-1}, i_{-2}, \ldots, i_{-k}, i''_{-k-1}, \ldots)} - 1 \right|$$

then $\sum_{k=1}^{\infty} \delta_k < \infty$.

Theorem 6.1 *There exists one and only one measure.μ on the Borel σ-algebra of the space M such that*
1.) μ is invariant under the shift;
2.) for any semi-infinite sequence $\ldots, i_{m-2}, i_{m-1}, i_m$ such that $a_{i_{k-1} i_k} = 1$, $k \le m$ the limit

$$\lim_{k \to \infty} \mu(i_m \mid i_{m-1}, i_{m-2}, \ldots, i_{m-k})$$

exists and equals to $p(i_m \mid i_{m-1}, i_{m-2}, \ldots)$.

Before going to the proof of this theorem let us explain the meaning of $\mathbf{i_3}$). Fix any admissible semi-infinite sequence $\ldots, i_{m-2}^0, i_{m-1}^0, i_m^0$ (i.e. $a_{i_{k-1}^0, i_k^0} = 1$) and consider the probability distribution P on 'future words', i.e.

$$
\begin{aligned}
P(i_{m+1}, i_{m+2}, \ldots, i_{m+k}) \; = \; & p(i_{m+1} \mid i_m^0, i_{m-1}^0, i_{m-2}^0, \ldots) \cdot \\
& p(i_{m+2} \mid i_{m+1}, i_m^0, i_{m-1}^0, i_{m-2}^0, \ldots) \cdots \\
& p(i_{m+k} \mid i_{m+k-1}, \ldots, i_{m+1}, i_m^0, i_{m-1}^0, \ldots).
\end{aligned}
$$

Certainly P depends on $\ldots, i_{m-2}^0, i_{m-1}^0, i_m^0$. The condition $\mathbf{i_3}$) implies that for different pasts $\ldots, i_{m-2}^0, i_{m-1}^0, i_m^0$ the probability distributions are equivalent. Indeed,

$$
\frac{P\left(i_{m+1}, i_{m+2}, \ldots, i_{m+k} \mid i_m^0, i_{m-1}^0, \ldots, i_{m-s}^0, \ldots\right)}{P\left(i_{m+1}, i_{m+2}, \ldots, i_{m+k} \mid i_m^1, i_{m-1}^1, \ldots, i_{m-s}^1, \ldots\right)} \le \exp\left\{\sum_{n=1}^{k} \delta_k\right\} \le \text{const}.
$$

Let us show that at least one μ with the needed properties exists. Fix an arbitrary admissible sequence $\{i_m^0\}_{-\infty}^{\infty}$ and for every s, $-\infty < s < \infty$ take the measure $P_s^{(0)}$ for which

$$P_s^{(0)}(\ldots, i_{s-m}^0, \ldots, i_s^0) = 1$$

$$P_s^{(0)}(i_{s+1}, \ldots, i_{s+t}) = \prod_{r=s+1}^{s+t} p(i_r \mid i_{r-1}, \ldots, i_{s+1}, i_s^0, i_{s-1}^0, \ldots).$$

Since M is compact in the topology of direct product, the space of probability measures on the Borel σ-algebra of M is weakly compact. Therefore for some subsequence $\{-t_j\}$, $t_j \to \infty$ as $j \to \infty$

$$P_{-t_j}^{(0)} \to P^{(0)}.$$

Our first claim says that for $P^{(0)}$

$$\lim_{n \to \infty} P^{(0)}(i_s \mid i_{s-1}, \ldots, i_{s-n}) = p(i_s \mid i_{s-1}, \ldots, i_{s-n}, \ldots) \qquad (6.1)$$

for any admissible sequence $i_s, i_{s-1}, \ldots, i_{s-n}, \ldots$.
    Indeed,

$$P^{(0)}(i_s \mid i_{s-1}, \ldots, i_{s-n}) = \lim_{j \to \infty} \frac{P_{-t_j}^{(0)}(i_s, i_{s-1}, \ldots, i_{s-n})}{P_{-t_j}^{(0)}(i_{s-1}, \ldots, i_{s-n})}.$$

But for admissible words

$$P^{(0)}_{-t_j}(i_s, i_{s-1}, \ldots, i_{s-n}) = \sum_{i'_{s-n-1}, \ldots, i'_{-t_j+1}}$$

$$\prod_{m=-t_j+1}^{s} p(i_m \mid i_{m-1}, \ldots, i_{s-n}, i'_{s-n-1}, \ldots, i'_{-t_j+1}, i'_{-t_j}, i'_{-t_j-1}, \ldots).$$

An analogous sum can be written for $P^{(0)}_{-t_j}(i_{s-1}, \ldots, i_{s-n})$.

Each term in the first sum has an extra factor which differs from $p(i_s \mid i_{s-1}, \ldots, i_{s-n}, \ldots)$ by an error which does not exceed δ_n. Letting $j \to \infty$ and then $n \to \infty$ we get (6.1).

Now we shall show that the measure satisfying (6.1) is unique. Our argument is based on the following lemma.

Lemma 6.2 *Let P be a probability measure defined on the Borel σ-algebra of M for which (6.1) holds. Then for $\varepsilon > 0$ one can find $m(\varepsilon) = m$, $t(\varepsilon) = t$ with $m(\varepsilon) \to \infty$, $t(\varepsilon) \to \infty$ as $\varepsilon \to 0$, and a probability distribution $\bar{P}^{(0,m)}$ on the words $\{i_{-m}, \ldots, i_m\}$ such that for any conditional distribution*

$$P_{i-t-1, i-t-2, \ldots}(i_{-m}, \ldots, i_m) = P(i_{-m}, \ldots, i_m \mid i_{-t-1}, i_{-t-2}, \ldots)$$

induced by P

$$\mathrm{Var}(\bar{P}^{(0,m)}, P_{i-t-1, i-t-2, \ldots}) \le \varepsilon.$$

Here Var *is the distance in variation between probability distributions.*

Proof. Assuming that $m(\varepsilon)$ is chosen take $t = t(\varepsilon) = q(\varepsilon)(2m(\varepsilon)+1) + m(\varepsilon)$ where $q(\varepsilon)$ is an integer to be chosen later. Write down the conditional probability

$$P(i_m, i_{m-1}, \ldots, i_{-t} \mid i_{-t-1}, \ldots) = \prod p(i_s \mid i_{s-1}, i_{s-2}, \ldots)$$

and replace it by a conditional probability of a Markov chain with memory $(2m+1)$ as follows:

$$P(i_m, i_{m-1}, \ldots, i_{-t} \mid i_{-t-1}, \ldots) = \prod_{l=1}^{q(\varepsilon)} \prod_{-t(\varepsilon)+l(2m+1)<s \le -t(\varepsilon)+(l+1)(2m+1)}$$

$$p(i_s \mid i_{s-1}, \ldots, i_{-t(\varepsilon)+(l-1)(2m+1)+1}, i_{-t(\varepsilon)+(l-1)(2m+1)}, j_1, j_2, \ldots) \exp\{\rho\}.$$

Here $j_1, j_2, \ldots$ is an arbitrary admissible sequence and ρ is an error. It follows easily from the condition i_3) that

$$|\rho| \le q(\varepsilon) \sum_{s=2m+1}^{4m+1} \delta_s.$$

Since $\sum \delta_s < \infty$ we can find for any $q = q(\varepsilon)$ so large m that

$$q(\varepsilon) \sum_{s=2m+1}^{4m+1} \delta_s \le \varepsilon.$$

Therefore if we put

$$\bar{P}(i_m, i_{m-1}, \ldots, i_{-t}) = \prod_{l=1}^{q(\varepsilon)} \prod_{-t(\varepsilon)+l(2m+1)<s\le -t(\varepsilon)+(l+1)(2m+1)}$$
$$p(i_s \mid i_{s-1}, \ldots, i_{-t(\varepsilon)+(l-1)(2m+1)+1}, i_{-t(\varepsilon)+(l-1)(2m+1)}, j_1, j_2, \ldots),$$

then $\operatorname{Var}\left(\bar{P}, P_{i_{-t-1}, i_{-t-2}, \ldots}\right) \le \frac{\varepsilon}{2}$.

The probability measure $\bar{P}$ corresponds to the homogeneous Markov chain of memory $(2m + 1)$. Its conditional probabilities are

$$\bar{P}(i_m, \ldots, i_{-m} \mid i_{-m-1}, \ldots, i_{-3m-1}) =$$
$$\prod p(i_s \mid i_{s-1}, \ldots, i_{-3m-1}, j_1, j_2, \ldots).$$

Introduce a linear operator Π acting on the space of probability distributions on the words $(i_m, \ldots, i_{-m})$ generated by the conditional probabilities $\bar{P}$. It has the following important property:

$$\frac{\bar{P}(i_m, \ldots, i_{-m} \mid i_{-m-1}, \ldots, i_{-3m-1})}{\bar{P}(i_m, \ldots, i_{-m} \mid i'_{-m-1}, \ldots, i'_{-3m-1})} \ge \text{const} = C_0 > 0,$$

where C_0 does not depend on m. Take as $\bar{P}^{(0,m)}$ stationary distribution of this Markov chain. It follows from the usual ergodic theorem for Markov chains that

$$\operatorname{Var}\left(\bar{P}^{(0,m)}, \Pi P\right) \le (1 - C_0)\operatorname{Var}\left(\bar{P}^{(0,m)}, P\right).$$

Using this inequality $q(\varepsilon)$ times, where $2(1 - C_0)^{q(\varepsilon)} < \frac{\varepsilon}{2}$, we immediately get

$$\operatorname{Var}\left(\bar{P}, \bar{P}^{(0,m)}\right) \le \frac{\varepsilon}{2}.$$

This gives the statement of the lemma.

Now we can prove uniqueness of probability distributions satisfying (6.1). Suppose that there exist two such distributions P, Q. Then for some $\alpha > 0$ and all sufficiently large m we have $\operatorname{Var}(P_{-m,m}, Q_{-m,m}) \ge \alpha$ where $P_{-m,m}$, $Q_{-m,m}$ are induced probability distributions on the words $(i_{-m}, \ldots, i_m)$. But we come to a contradiction since according to the lemma

$$\operatorname{Var}\left(P_{-m,m}, \bar{P}^{(0,m)}\right) \le \varepsilon, \quad \operatorname{Var}\left(Q_{-m,m}, \bar{P}^{(0,m)}\right) \le \varepsilon.$$

Thus we see that P is unique. Since the shift of the measure satisfying (6.1) is also a measure satisfying (6.1) we see that the shift of P coincides with P, i.e. P is invariant under the shift.

Remark 6.3 *In the previous proof the condition* i_3) *plays the most important role. It is closely connected with Ruelle's criterium of the absence of phase transitions in one-dimensional lattice models of statistical mechanics. Probably it is optimal in the following sense. For many sequences* δ_n, $\sum \delta_n = \infty$ *one can construct conditional probabilities p for which there exist several probability distributions P satisfying (6.1).*

It would be interesting to study this problem in detail.

Theorem 6.1 has some applications to the theory of expanding maps. Let M be a compact closed metric space and T a continuous map of M onto itself.

Definition 6.4 *T is called* expanding *if for some* $\delta > 0$ *and* $\lambda > 1$

$$d(Tx, Ty) \geq \lambda d(x, y)$$

provided that $d(x, y) \leq \delta$.

The first result in the theory of expanding maps is the existence of a natural invariant measure proven under very mild additional conditions. We shall discuss related problems in the simplest one-dimensional case. Let $T : x \to f(x)$, $x \in [0, 1]$ where f is a piece-wise continuous function such that on any component of continuity $f'(x) \geq \lambda > 1$ and f' satisfies a Hölder condition with Hölder exponent γ, i.e.

$$|f'(x_1) - f'(x_2)| \leq C_1 |x_1 - x_2|^\gamma$$

for some constant C_1. Additionally we assume that if $\Delta_1, \Delta_2, \ldots, \Delta_r$ are components of continuity then $T\Delta_i = [0, 1]$. Below we prove the following theorem

Theorem 6.5 *T has an absolutely continuous invariant measure.*

There is a short proof of this theorem which does not give too much information about this measure. We shall present here a longer proof which reduces Theorem 6.5 to Theorem 6.1 and shows deep connections with statistical mechanics of lattice systems and Gibbs random fields.

We begin with the construction of a 'symbolic representation' of T. Take any point x such that $T^k x$, $k = 0, 1, \ldots$ is not a boundary point of any Δ_i, $i = 1, \ldots, r$. Consider for such x the sequence of inclusions $T^k x \in C_{i_k}$, $k = 0, 1, \ldots$. The sequence of indices $\{i_0, i_1, i_2, \ldots\}$ is defined in a unique way. It is called *symbolic representation* of x. The point Tx has the symbolic representation $\{i_1, i_2, \ldots\}$. We can also write

$$x = \Delta_{i_0} \cap T^{-1}\Delta_{i_1} \cap \ldots \cap T^{-n}\Delta_{i_n} \cap \ldots.$$

Now we can take an arbitrary sequence $i = \{i_0, i_1, \ldots, i_n, \ldots\}$ and consider the intersection

$$\bigcap_{n=0}^{\infty} T^{-n}\Delta_{i_n} = x(i).$$

Since the Δ_j are closed and $T\Delta_j = [0,1]$ this intersection is non-empty. Points which pass through the boundary $\cup_{j=1}^r \partial \Delta_j$ have a non-unique representation.

Denote by Ω the space of all sequences $i = \{i_0, i_1, \ldots, i_n, \ldots\}$, $i_n = 1, 2, \ldots, r$. We have a mapping $\varphi : i \mapsto x(i)$ which is not one-to-one only on a countable subset. If S is the shift in Ω, i.e. $Si = \{i_1, \ldots, i_n, \ldots\}$ then $\varphi S = T\varphi$.

Take an arbitrary semi-infinite sequence

$$i^- = \{\ldots, i_{-n}, \ldots, i_{-2}, i_{-1}, i_0\}.$$

Lemma 6.6 *There exists the following limit*

$$p(i_0 \mid i_{-1}, \ldots, i_{-n}, \ldots)$$
$$= \lim_{n\to\infty} \frac{l\left(\Delta_{i_{-n}} \cap T^{-1}\Delta_{i_{-n+1}} \cap \ldots \cap T^{-n+1}\Delta_{i_{-1}} \cap T^{-n}\Delta_{i_0}\right)}{l\left(\Delta_{i_{-n}} \cap T^{-1}\Delta_{i_{-n+1}} \cap \ldots \cap T^{-n+1}\Delta_{i_{-1}}\right)},$$

where l is the length.

The limiting function p satisfies the conditions $\mathbf{i_1}$), $\mathbf{i_2}$), $\mathbf{i_3}$) *with* $\delta_n = \text{const } \rho^n$ *for some* $\rho < 1$.

Proof. Denote

$$l_n = l(i_0 \mid i_{-1}, \ldots, i_{-n}) = \frac{l\left(\cap_{k=0}^n T^{-k}\Delta_{i_{-n+k}}\right)}{l\left(\cap_{k=0}^{n-1} T^{-k}\Delta_{i_{-n+k}}\right)}.$$

Consider the ratio

$$\frac{l_{n+1}}{l_n} = \frac{l\left(\cap_{k=0}^{n+1} T^{-k}\Delta_{i_{-n-1+k}}\right)}{l\left(\cap_{k=0}^n T^{-k}\Delta_{i_{-n+k}}\right)} \cdot \frac{l\left(\cap_{k=0}^{n-1} T^{-k}\Delta_{i_{-n+k}}\right)}{l\left(\cap_{k=0}^n T^{-k}\Delta_{i_{-n-1+k}}\right)}.$$

We also have

$$T\left(\cap_{k=0}^{n+1} T^{-k}\Delta_{i_{-n-1+k}}\right) = \bigcap_{k=0}^n T^{-k}\Delta_{i_{-n+k}},$$

$$T\left(\cap_{k=0}^n T^{-k}\Delta_{i_{-n-1+k}}\right) = \bigcap_{k=0}^{n-1} T^{-k}\Delta_{i_{-n+k}}.$$

Since f is continuous on each Δ_i

$$l\left(\bigcap_{k=0}^n T^{-k}\Delta_{i_{-n+k}}\right) = \int_{\cap_{k=0}^{n+1} T^{-k}\Delta_{i_{-n-1+k}}} f'(x)\,dx$$
$$= f'(y_1)\, l\left(\bigcap_{k=0}^{n+1} T^{-k}\Delta_{i_{-n-1+k}}\right)$$

and

$$l\left(\bigcap_{k=0}^{n-1}T^{-k}\Delta_{i-n+k}\right) = \int_{\bigcap_{k=0}^{n}T^{-k}\Delta_{i-n-1+k}} f'(x)dx$$

$$= f'(y_2)l\left(\bigcap_{k=0}^{n}T^{-k}\Delta_{i-n-1+k}\right).$$

Here y_1, y_2 are some intermediate points.

Therefore $l_{n+1}/l_n = f'(y_1)/f'(y_2)$. Since $\text{dist}(y_1,y_2) \leq \text{const}\,\rho^n$ for some $\rho < 1$ and

$$\frac{f'(y_1)}{f'(y_2)} = 1 + \frac{f'(y_1) - f'(y_2)}{f'(y_2)},$$

we conclude that $|f'(y_1) - f'(y_2)| \leq \text{const}\,\rho^{n\gamma}$ and

$$\left|\frac{l_{n+1}}{l_n} - 1\right| \leq \text{const}\,\rho_1^n$$

for some $\rho_1 < 1$. These inequalities show that $\lim_{n\to\infty} l_n$ exists, and the lemma is proven.

Now we can use Theorem 6.1 which gives the existence of measure μ_0 on the Borel σ-algebra of the space Ω having p as one-sided conditional probabilities.

Using the mapping φ we can construct a measure μ on $[0,1]$. This measure

1. is invariant under T (that follows from the invariance of μ_0 under S)

2. is absolutely continuous with respect to the Lebesgue measure.

One can easily show that T is an exact endomorphism with exponential decay of correlations.

This exposition follows my book [1].

Bibliography

[1] Ya. G. Sinai: Topics in Ergodic Theory. Princeton University Press. Princeton, New Jersey 1994

Chapter 7

Liouville Surfaces

Liouville surfaces are Riemannian surfaces equipped with Liouville metrics. These metrics can be on spheres or on tori. We shall consider the case of tori.

Let the torus Q be realized as the unit square with the corresponding boundary conditions and coordinates q_1, q_2. The Liouville metric is given by two functions $U_1(q_1)$, $U_2(q_2)$ and has the form

$$dq^2 = (U_1(q_1) - U_2(q_2))(dq_1^2 + dq_2^2).$$

Naturally $U_1(q_1) > U_2(q_2)$ everywhere. The main peculiar property of Liouville surfaces is that the corresponding geodesic flow is integrable, i.e. its phase space can be decomposed into invariant two-dimensional tori and a subset of measure zero. This is remarkable because the geodesic flow has no symmetry and only an additional first integral.

The Hamiltonian system for the geodesic flow has the form

$$\dot{p}_1 = \frac{H(\mathbf{p}, \mathbf{q})}{U_1(q_1) - U_2(q_2)} \frac{dU_1}{dq_1}, \qquad \dot{p}_2 = -\frac{H(\mathbf{p}, \mathbf{q})}{U_1(q_1) - U_2(q_2)} \frac{dU_2}{dq_2}$$

$$\dot{q}_1 = \frac{2p_1}{U_1(q_1) - U_2(q_2)}, \qquad \dot{q}_2 = \frac{2p_2}{U_1(q_1) - U_2(q_2)}$$

with Hamiltonian function

$$H(\mathbf{p}, \mathbf{q}) = \frac{p_1^2 + p_2^2}{U_1(q_1) - U_2(q_2)}$$

and first integral

$$S(\mathbf{p}, \mathbf{q}) = \frac{U_2(q_2)}{U_1(q_1) - U_2(q_2)} p_1^2 + \frac{U_1(q_1)}{U_1(q_1) - U_2(q_2)} p_2^2.$$

It is conceivable that for two-dimensional surfaces every generic integrable metric can be reduced to a Liouville surface.

We shall assume that each of the functions U_1, U_2 is C^∞-smooth and has two non-degenerate critical points. In this case the structure of geodesics is especially simple. The geodesics can be of one of three types:

1. Vertical-like geodesics. There are two vertical-like curves and all geodesics are between the two curves and are tangent to them (Fig. 7.1).

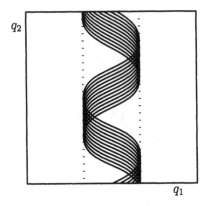

Figure 7.1: Geodesic on an invariant torus extending in the q_1-direction

2. Geodesics which are everywhere dense on the torus (Fig. 7.2).

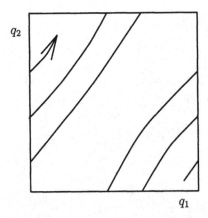

Figure 7.2: Geodesic on an invariant torus projecting diffeomorphically to Q

3. Horizontal-like geodesics (Fig. 7.3)

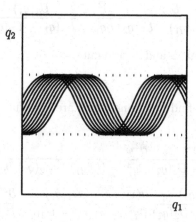

Figure 7.3: Geodesic on an invariant torus extending in the q_1-direction

These three types correspond to three intervals of values of S. With $S = cE$ one has the correspondence (Fig. 7.4)

1. $c_2 < c < c_1$

2. $c_3 < c < c_2$

3. $c_4 < c < c_3$.

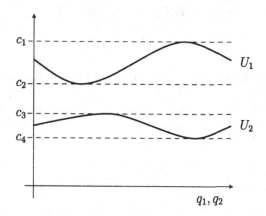

Figure 7.4: Parameter ranges for the first integral S

We shall study spectral properties of the Laplace-Beltrami operators on Liouville surfaces. These operators have the form

$$\Delta = -\frac{1}{U_1(q_1) - U_2(q_2)} \left(\frac{\partial^2}{\partial q_1^2} + \frac{\partial^2}{\partial q_2^2} \right).$$

Consider the quantum analog of the first integral S

$$\hat{S} = -\frac{U_2(q_2)}{U_1(q_1) - U_2(q_2)}\frac{\partial^2}{\partial q_1^2} - \frac{U_1(q_1)}{U_1(q_1) - U_2(q_2)}\frac{\partial^2}{\partial q_2^2}.$$

The key remark says that Δ and $\hat{S}$ commute:

$$\Delta\hat{S} = \Delta\hat{S}.$$

Therefore we can try to find functions which are eigenfunctions of both operators simultaneously:

$$-\frac{1}{U_1(q_1) - U_2(q_2)}\frac{\partial^2\psi}{\partial q_1^2} - \frac{1}{U_1(q_1) - U_2(q_2)}\frac{\partial^2\psi}{\partial q_2^2} = E^1\psi$$

$$-\frac{U_2(q_2)}{U_1(q_1) - U_2(q_2)}\frac{\partial^2\psi}{\partial q_1^2} - \frac{U_1(q_1)}{U_1(q_1) - U_2(q_2)}\frac{\partial^2\psi}{\partial q_2^2} = E^2\psi.$$

This system is equivalent to the following system.

$$\frac{\partial^2\psi}{\partial q_1^2} + \left(E^1U_1(q_1) - E^2\right)\psi = 0$$

$$\frac{\partial^2\psi}{\partial q_2^2} + \left(E^2 - E^1U_2(q_2)\right)\psi = 0.$$

It shows that eigenfunctions ψ have the form $\psi(q_1, q_2) = \psi^1(q_1) \cdot \psi^2(q_2)$ and

$$\frac{d^2\psi^1}{dx^2} + \left(E^1U_1(x) - E^2\right)\psi^1 = 0$$

$$\frac{d^2\psi^2}{dx^2} + \left(E^2 - E^1U_2(x)\right)\psi^2 = 0. \tag{7.1}$$

It is easy to see that the system (7.1) cannot have periodic solutions if $E^1U_1 - E^2 < 0$ or $E^2 - E^1U_2 < 0$.

Our main purpose is to analyse asymptotical properties of eigenvalues of $-\Delta$. Let us recall the first general result (Weyl's asymptotics): If $N(x)$ is the number of eigenvalues of $-\Delta$ less than x then

$$N(x) = \frac{\mathrm{Area}(Q)}{4\pi}x + n(x)$$

and $\lim_{x \to \infty} \frac{n(x)}{x} = 0$. We shall study $n(x)$ for Liouville surfaces and we shall show that $n(x)$ has a number-theoretical nature. This is specific for integrable surfaces, i.e. for surfaces where the geodesic flow is integrable.

Introduce the following functions:

$$F_1(c) = \begin{cases} \int_0^1 (U_1(q_1) - c)^{\frac{1}{2}}\,dq_1 & \text{for } c_4 \leq c \leq c_2 \\ \int_{q_1'(c)}^{q_1''(c)} (U_1(q_1) - c)^{\frac{1}{2}}\,dq_1 & \text{for } c_2 \leq c \leq c_1 \end{cases}$$

$$F_2(c) = \begin{cases} \int_{q_2'(c)}^{q_2''(c)} (c - U_2(q_2))^{\frac{1}{2}} dq_2 & \text{for } c_4 \leq c \leq c_3 \\ \int_0^1 (c - U_2(q_2))^{\frac{1}{2}} dq_2 & \text{for } c_3 \leq c \leq c_1 \end{cases}.$$

Here $q_1'(c), q_1''(c)$ (resp. $q_2'(c), q_2''(c)$) are solutions of the equations

$$U_1(q_1) - c = 0 \quad (\text{resp. } U_2(q_2) - c = 0).$$

The main interpretation:

Let $M_{1,c} = \{x \in M \mid H(x) = 1, S(x) = c\}$. Then

1. $c_2 < c < c_1$: $M_{1,c}$ consists of two invariant tori; their projections to the configuration space Q coincide and look like the vertical strip $q_1'(c) \leq q_1 \leq q_1''(c)$.

2. $c_3 < c < c_2$; $M_{1,c}$ consists of four invariant tori. Each of them is projected onto the whole torus Q.

3. $c_4 < c < c_3$; $M_{1,c}$ consists of two invariant tori whose projections to Q are horizontal strips $q_2'(c) \leq q_2 \leq q_2''(c)$.

The main result of the semi-classical analysis is the following theorem.

Theorem 7.1 *All eigenfunctions of the Laplacian $-\Delta$ and $\hat{S}$ can be parametrized by pairs $m = (m_1, m_2)$ where $m_i \geq 0$ are integers, $i = 1, 2$. The corresponding pairs of eigenvalues are solutions of the equation*

$$\Phi_0(L, c) + \Phi_1(L, c) + \Phi_2(L, c) = 2\pi \left(\left[\frac{m_1 + 1}{2} \right], \left[\frac{m_2 + 1}{2} \right] \right).$$

Here the eigenvalue $E^1 = L^2$, the eigenvalue $E^2 = L^2 \cdot c$, and

i) $\Phi_0(L, c) = L(F_1(c), F_2(c))$,

ii) $\Phi_1(L, c) = (\varphi_{\pm}^1(L, c), \varphi_{\pm}^2(L, c))$. *The choice of signs is determined by the parity of m_1, m_2: Even numbers correspond to $+$, odd numbers correspond to $-$.*

Even more,

$$\begin{array}{ll} |\varphi_{\pm}^1(L, c) \pm \frac{1}{2}\pi| \leq \text{const} L^{-2/3} \ln L & c_2 + \text{const} L^{-2/3} \leq c \leq c_1; \\ |\varphi_{\pm}^1(L, c)| \leq \text{const} L^{-2/3} \ln L & c_4 \leq c \leq c_2 - \text{const} L^{-2/3}; \\ |\varphi_{\pm}^1(L, c)| \leq \text{const} & \text{in all other cases.} \end{array}$$

$$\begin{array}{ll} |\varphi_{\pm}^2(L, c)| \leq \text{const} L^{-2/3} \ln L & c_3 + \text{const} L^{-2/3} \leq c \leq c_1; \\ |\varphi_{\pm}^2(L, c) \pm \frac{1}{2}\pi| \leq \text{const} L^{-2/3} \ln L & c_4 \leq c \leq c_3 - \text{const} L^{-2/3}; \\ |\varphi_{\pm}^2(L, c)| \leq \text{const} & \text{in all other cases.} \end{array}$$

iii) $\Phi_2(L, c) \leq \text{const} L^{-2/3} \ln L$ *for all $c \in [c_4, c_1]$.*

This theorem is a refined version of the Bohr-Sommerfeld-Keller-Maslov quantization rule. Φ_1 is connected with the Maslov index, Φ_2 is a remainder term.

Using this theorem we can give a geometric description of the function $N(x)$.
Consider in the positive quadrant the curve

$$\Gamma = (F_1(c), F_2(c)), \qquad c_4 < c < c_1.$$

In polar coordinates $\Gamma = \{(\rho, \alpha) \mid \rho = G(\alpha), 0 \le \alpha \le \frac{1}{2}\pi\}$ and

$$\tan \alpha = \frac{F_2(c(\alpha))}{F_1(c(\alpha))}, \quad G(\alpha) = \sqrt{F_1(c(\alpha))^2 + F_2(c(\alpha))^2}.$$

The values c_4, c_3, c_2, c_1 correspond to the angles:

$$\alpha_0 = \alpha(c_4) = 0, \qquad \alpha_1 = \alpha(c_3) = \arctan \frac{F_2(c_3)}{F_1(c_3)},$$

$$\alpha_2 = \alpha(c_2) = \arctan \frac{F_2(c_2)}{F_1(c_2)}, \qquad \alpha_3 = \frac{\pi}{2}.$$

The rays $\alpha = \alpha_1$ and $\alpha = \alpha_2$ decompose the quadrant into three sectors

$$A_i = \{(\rho, \alpha) \mid \alpha_{i-1} \le \alpha \le \alpha_i, \rho > 0\}, \qquad i = 1, 2, 3.$$

Introduce also the lattices

$$\begin{aligned}
\mathcal{L}_1 &= \{(2\pi k_1, \pi(k_2 + \tfrac{1}{2})) \mid (k_1, k_2) \in \mathbb{Z}^2\}, \\
\mathcal{L}_2 &= \{(2\pi k_1, 2\pi k_2) \mid (k_1, k_2) \in \mathbb{Z}^2\}, \\
\mathcal{L}_3 &= \{(\pi(k_1 + \tfrac{1}{2}), 2\pi k_2) \mid (k_1, k_2) \in \mathbb{Z}^2\}.
\end{aligned}$$

Now let $\mathcal{D}$ be a subset of the first quadrant bounded by Γ and two straight segments
on the lines $\alpha = 0$ and $\alpha = \pi/2$. Denote by $R\mathcal{D}$ the image of $\mathcal{D}$ under the scaling
with the coefficient R. Let also $N_i(R)$ be the number of the points of $\mathcal{L}_i$ inside
$A_i \cap R\mathcal{D}$.

Theorem 7.2 *For generic Liouville surfaces*

$$N(R^2) = 2N_1(R) + 4N_2(R) + 2N_3(R) + R^{\frac{1}{2}}\Theta_1(R),$$

*where $\Theta_1(R)$ is a quasi-periodic function of Besicovich class $B^1([0, \infty))$ equivalent
to zero.*

Theorem 7.2 shows that $N(R^2)$ can be represented up to some error as a sum of
points of lattices in expanding domains.

Bibliography

[1] D.V. Kosygin, A.A. Minasov, Ya.G. Sinai: Statistical properties of the Laplace-Beltrami operator on Liouville surfaces. Uspekhi Mat. Nauk **48**, 3–130 (1993)

[2] P. Bleher, D.V. Kosygin, Ya.G. Sinai: Distribution of energy levels of quantum free particle on the Liouville surface and trace formulae. Commun. Math. Phys. **170** 375–403 (1995)

Participants

1	Arians, Silke	Aachen	silke@iram.rwth-aachen.de
2	Bäcker, Arnd	Ulm	baec@physik.uni-ulm.de
3	Baladi, Viviane	Genf	baladi@sc2a.unige.ch
4	Bauer, Gernot	München	bauer@rz.mathematik.uni-muenchen.de
5	Berndl, Karin	München	berndl@rz.mathematik.uni-muenchen.de
6	Chang, Cheng-Hung	Clausthal	ch.chang@tu-clausthal.de
7	Dierkes, Tanja	Berlin	dierkes@marie.physik.tu-berlin.de
8	Fassomytakis, Jannis	München	fassomyt@rz.mathematik.uni-muenchen.de
9	Geyer, Lukas	Berlin	geyer@math.tu-berlin.de
10	von der Heyden, Arnd	Aachen	arnd@iram.rwth-aachen.de
11	Heisé, Hella	Berlin	
12	Herbig, Hans-Christian	Berlin	
13	Hövermann, Frank	Berlin	hoeverma@math.tu-berlin.de
14	Jung, Wolf	Aachen	jung@iram.rwth-aachen.de
15	Knauf, Andreas	Berlin	knauf@math.tu-berlin.de
16	Lorbeer, Boris	Berlin	lorbeer@math.tu-berlin.de
17	Nobbe, Burkhard	Berlin	nobbe@math.tu-berlin.de
18	Reuß, Alexander	Marburg	
19	Schubert, Roman	Ulm	schub@physik.uni-ulm.de
20	Sinai, Yakov	Moskau	
		Princeton	sinai@math.Princeton.EDU
21	Spoddeck, Heike	Berlin	spoddeck@pils.physik.tu-berlin.de
22	Swat, Maciej	Berlin	maciej@emmi.physik.tu-berlin.de
23	Teschel, Gerald	Aachen	gerald@iram.rwth-aachen.de
24	Teufel, Stefan	München	
25	Verbitski, Evgueni	Groningen	verbitski@math.rug.nl
26	Ziewer, Lukas	Bern	ziewer@math-stat.unibe.ch

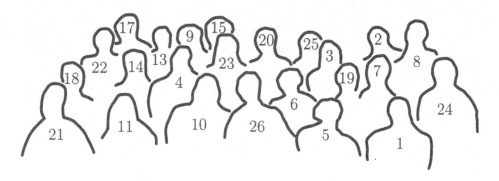

Participants of the seminar

Additional Talks

V. Baladi	Dynamical Zeta Functions for Pedestrians
A. Knauf	Topological Dynamics
Ya. Sinai	Hyperbolic Dynamical Systems
Ya. Sinai	Geodesic Flows on Compact Manifolds of Negative Curvature
A. Bäcker	Mode Fluctuations as Fingerprints of Chaotic and Non-Chaotic Systems
A. Bäcker	Symbolic Dynamics and Periodic Orbits for the Cardoid Billiard
C-H. Chang	Dynamical Interpretation of Selberg Zeta Function
F. Hövermann	Inverse Spectral Theory for 1-D Schrödinger Operators with Periodic Potential
W. Jung	Inverse Scattering for the Dirac Equation
B. Nobbe	2-D Diffusion
G. Teschl	Inverse Theory and Trace Formulas for Discrete 1-D Schrödinger Equation
A. Verbitzki	Generalized Entropies

Index